BestMasters

Mit **„BestMasters“** zeichnet Springer die besten Masterarbeiten aus, die an renommierten Hochschulen in Deutschland, Österreich und der Schweiz entstanden sind. Die mit Höchstnote ausgezeichneten Arbeiten wurden durch Gutachter zur Veröffentlichung empfohlen und behandeln aktuelle Themen aus unterschiedlichen Fachgebieten der Naturwissenschaften, Psychologie, Sozialwissenschaften, Technik und Wirtschaftswissenschaften. Die Reihe wendet sich an Praktiker und Wissenschaftler gleichermaßen und soll insbesondere auch Nachwuchswissenschaftlern Orientierung geben.

Springer awards **“BestMasters”** to the best master’s theses which have been completed at renowned Universities in Germany, Austria, and Switzerland. The studies received highest marks and were recommended for publication by supervisors. They address current issues from various fields of research in natural sciences, psychology, social sciences, technology, and economics. The series addresses practitioners as well as scientists and, in particular, offers guidance for early stage researchers.

Maren Stümke

Identifying and Overcoming Student's Misconceptions in Chemistry

An Intelligent Tutoring System to Teach Chemical Equations with Worked-out Analogical Examples

Springer Vieweg

Maren Stümke
Otto-Friedrich-Universität Bamberg
Berlin, Germany

ISSN 2625-3577 ISSN 2625-3615 (electronic)
BestMasters
ISBN 978-3-658-51031-2 ISBN 978-3-658-51032-9 (eBook)
https://doi.org/10.1007/978-3-658-51032-9

This Springer Vieweg imprint is published by the registered company Springer Fachmedien Wiesbaden GmbH, part of Springer Nature.
The registered company address is: Abraham-Lincoln-Str. 46, 65189 Wiesbaden, Germany

Acknowledgements

I developed this master thesis in cooperation with the "Institut für Pädagogik in den Naturwissenschaften" in Kiel. The idea was to add a chemical system to the existing tutoring systems at the Chair of Cognitive Systems at the University of Bamberg. Redox reactions are particularly suitable for this. However, since I do not have the chemical expertise to implement a system to guide students in solving redox reactions, I would first like to thank S and G for their many chemical tips and explanations. Thanks also to S for providing materials such as textbooks and the Excel spreadsheet in the appendix.

S also supervised this thesis (especially the chemistry part) together with B (for the computer science part). Thank you both for the many questions you have answered over the last few months and thank you, B, for the regular meetings. These have helped me greatly organize my master's thesis over time.

I would like to thank A and S for participating in the interview and for giving detailed and helpful feedback. As mentioned in the thesis, this feedback forms a good basis for considering future improvements to the system.

A special thanks to T, A, M, and F for their thorough proofreading and many helpful comments. Thank you, M, for getting me out of bed every morning these past weeks. Thanks to V for helping me create the *Chemielon*. Finally, I would like to thank T and N for their support in printing and submitting the thesis.

Abstract (english)

Intelligent Tutoring Systems (ITS) aim to provide students with comparable instructional advantages to that of a human tutor by responding to individual needs and providing immediate feedback. Originally dating back to the 1980s, they have proven useful in different academic and educational contexts, especially in STEM-related domains (VanLehn, 2011). However, an ITS helping students to understand the underlying system of building and solving chemical formulae is lacking although many students struggle with these concepts (Taskin and Bernholt, 2012). Consequently, this Master Thesis provides the implementation and some first testing of an ITS for chemical formulae, especially redox reactions. It follows the classic architecture of an ITS with a background knowledge comprising the rules of deriving equations, a student module identifying errors or misconceptions in students' answers, a simple interface, and a feedback module where structurally analogous examples are selected. We chose analogies as a feedback method as they enable students to not only solve the example at hand but also grasp the underlying schemata. This is essential to truly understand chemical formulae. The ITS was tested in a qualitative evaluation study and proved to be processing reliably and being reasonably easy to understand.

Abstract (deutsch)

Intelligente Tutorensysteme (ITS) zielen darauf ab, den Studierenden vergleichbare Vorteile wie ein menschlicher Tutor zu bieten, indem sie auf die individuellen Bedürfnisse eingehen und sofortiges Feedback geben. Seit den Anfängen solcher Systeme in den 1980er Jahren haben sie sich in verschiedenen akademischen und pädagogischen Kontexten als nützlich erwiesen, insbesondere in MINT-Fächern und -Bereichen (VanLehn, 2011). Bisher fehlt jedoch ein ITS, welches Schüler:innen hilft, die theoretischen Grundlagen und Konzeption chemischer Formeln zu verstehen. Da viele Schüler:innen aber Probleme mit dem Ausgleichen chemischer Formeln, gerade Redoxreaktionen, haben (Taskin and Bernholt, 2012), bietet diese Masterarbeit die Implementierung und erste Tests eines ITS für solche Formeln. Die Arbeit folgt dabei der klassischen Architektur eines ITS mit einem Modul für das Hintergrundwissen, welches die Regeln zur Herleitung von Gleichungen umfasst, einem Schüler:inen-Modul, das Fehler oder Missverständnisse in den Antworten der Schüler:innen identifiziert, einem Frontend und einem Feedbackmodul, in dem strukturell analoge Beispiele ausgewählt werden. Analogien werden als Rückmeldung genutzt, da sie nicht nur helfen, einzelne Beispiele zu lösen, sondern auch die dahinterstehenden abstrakten Zusammenhänge und Schemata zu erfassen. Dies ist für ein echtes Verständnis chemischer Formeln unerlässlich. Das ITS wurde in einer qualitativen Evaluationsstudie getestet und erwies sich als robust, zuverlässig und leicht verständlich.

Contents

List of Figures

List of Tables

1 Introduction

> "You see, wire telegraph is a kind of a very, very long cat. You pull his tail in New York and his head is meowing in Los Angeles. Do you understand this? And radio operates exactly the same way: you send signals here, they receive them there. The only difference is that there is no cat" (Albert Einstein).

The example above constitutes an analogy: The structure inherent to the sentences is used to explain a connection between different aspects. Although analogies are often used as a type of common-sense reasoning, to underline the understanding of the point one makes by comparing two objects or ideas (Pritchard, 2018), they are also useful for higher-order, deductive reasoning, for understanding abstract relations between objects.[1] Consider that in the quote above Albert Einstein uses an analogy to explain (at the time) a new and complicated technology which nevertheless seems perfectly easy to understand because of the analogy. Analogies of the form "A is to B as C is to D" are an essential mechanism in cognitive sciences, especially regarding learning, problem-solving, and transfer across domains. Analogical reasoning results in mental models which—in turn—may lead to the understanding of new domains. Reasoning by analogy thereby fosters learning: By understanding the connection drawn between different objects or domains, we transfer our knowledge from known examples onto an unknown domain—and learn something new.

Analogies have been successfully applied to many complex, abstract domains, including chemistry. Using analogies helps students to grasp abstract chemical con-

[1] A relation is simply any predicate taking more than one argument. An attribute (or object-attribute) captures the properties of an object. For example, $large(x)$ is an attribute as it only takes one argument while $larger(x, y)$ is a relation because it is a binary predicate (Gentner, 1983). Additionally, first-order relations take objects as arguments while higher-order relations take propositions as argument.

M. Stümke, *Identifying and Overcoming Student's Misconceptions in Chemistry*, BestMasters, https://doi.org/10.1007/978-3-658-51032-9_1

cepts better (Orgill and Bodner, 2004). They have also been used as the feedback mechanism within tutoring systems, for example, on written subtraction (Zeller, 2015).

In general, Intelligent Tutoring Systems (ITS) aim to provide students with comparable instructional advantages to that of a human tutor by responding to individual needs and providing immediate feedback. They have proven useful in different academic and educational contexts, especially in STEM-related domains (VanLehn, 2011). But, unlike analogies, they have rarely been applied to chemistry. Indeed, an ITS to help students understand the underlying system of building and solving chemical equations, especially redox reactions, is lacking even though many students struggle with these important concepts (Taskin and Bernholt, 2012). Tutoring systems are a very promising strategy for at least two reasons: Firstly, as chemistry is a STEM-related domain, it is likely that an ITS will yield good results here. Secondly, the acquisition of procedural knowledge tasks (such as solving redox reactions) requires identifying misconceptions and providing suitable feedback to overcome them.[2] An ITS—especially combined with analogies—provides this and is therefore advised: "One implication of this review [...] is that step-based tutoring systems should be used [...] in frequently offered or large enrollment STEM courses" (VanLehn, 2011, p. 213).

Consequently, this master thesis aims to provide the implementation and some initial testing of an ITS for chemical equations. It will follow the classic architecture of an ITS with a background knowledge module comprising both rules and facts to solve equations, a student module to identify errors and their underlying misconceptions, a (simple) interface, and a feedback module where structurally analogous worked-out examples are chosen. Therefore, if a student makes a mistake, the system shows a similarly structured formula and its correct solution. This way, the analogous solution can be transferred to the problem at hand to solve it correctly and understand the structure behind the construction of such equations. For example, if a task has been answered incorrectly, an ITS can give the solution to an analogous problem. The students would then have to recognize the analogy to apply the solution to their new task. Analogies are hence a very fruitful feedback method here: They potentially enable students to not only solve the example at hand but also grasp the underlying structure, as analogical reasoning results in the construction of mental models (Gentner and Smith, 2013).

[2] Misconceptions are psychologically plausible misunderstandings of the student that account for bugs manifested in the student's programs (Murray, 1987). Such misconceptions include overgeneralization, the inability to combine multiple constructs, or making incorrect analogies.

The larger research question is the following: Does analogical reasoning within an ITS help students to better understand the underlying structure of chemical equations and to solve them correctly? However, to fully answer the question, an empirical study with students is needed. As conducting such a study goes beyond the scope of this master thesis, I will aim to answer a second, smaller research question here: Is the Tutoring System easy to understand and use, so the system has the potential to help students solve chemical equations in the future? To answer this second research question, the ITS entailing analogical reasoning is not only implemented but also tested in a qualitative evaluation study using interview techniques to check if it can be processed reliably and is easy to understand.

I will lay out the work here by first summarizing the basic concepts of an ITS. In the third chapter, the feedback module of a tutoring system is described in more length, thereby shedding light on the decision to use analogies as a feedback method. These are then introduced in the fourth chapter along with the concept of algorithmic debugging I drew on. The fifth chapter explains the process of the implementation. This implementation is then evaluated and discussed, and possible future extensions of the system are outlined. Lastly, this thesis concludes with a summary of the findings.

Basic Concepts and Current Research on ITS 2

An ITS is an adaptive instructional system that incorporates both AI techniques and educational methods and is grounded in cognitive theories of learning and artificial intelligence (Rau et al., 2015). It includes a cognitive model of the students' problem-solving steps. This cognitive model allows one to infer which step the student is currently working on and to provide hints on how to solve this step or to provide adaptive error feedback. If a student makes a mistake indicative of a common misconception, the ITS provides feedback designed to challenge it.[1] Hence, an ITS not only avoids the one-to-many student-teacher model but also provides students with individualized, customized support tailored to their level of proficiency (Harsley et al., 2016).

The systems thereby provide highly developed instructional guidance on a one-to-one foundation aiming to be as good as that of a decent human tutor. Research shows that using these tutoring systems can make swift improvements and radically enhance a student's performance in precise areas and skills as they are tailored to the needs of the student. Indeed, Almasri et al. (2019) claim that the future of the development of the educational process lies in the development of ITS because they adapt to the intellectual levels of the student, are easy to use, and are available to everyone.

In this chapter, I will first define an ITS in more detail, explain its four core modules, and present some existing tutoring systems that relate to my research question. Again, this list of ITS reveals the lack of chemistry-related systems, especially systems focusing on equations.

[1] Some tutoring systems also adapt the selection of tutor problems to the individual's learning progress. However, as this is not a feature shared by all systems, I will not focus on this aspect here.

M. Stümke, *Identifying and Overcoming Student's Misconceptions in Chemistry*, BestMasters, https://doi.org/10.1007/978-3-658-51032-9_2

2.1 Defining an Intelligent Tutoring System

The idea of an ITS is to provide students with the same instructional advantages of a human tutor by responding to their individual needs to improve the quality of education. It is usually software aiming to provide immediate and customized instruction or feedback to learners so that learning is achieved efficiently using computing technologies. Such customized and immediate feedback is useful as the capacity of the working memory—engaged in remembering one's solution or learning by heart—is limited and prone to errors. Indeed, many errors learners make are due to failures of the working memory rather than failures of understanding (Anderson et al., 1985). Hence, enabling students to learn general problem-solving mechanisms is a great advantage.

Empirical evidence suggests that an ITS can live up to these expectations. For example, tutoring systems have outperformed students learning the task on their own (Anderson et al., 1985). Students learned given material four times faster with an ITS, and a review of experiments proved that an ITS—by comparison of effect sizes—is as effective as human tutoring (VanLehn, 2011).

Furthermore, an ITS is in essence interdisciplinary as the systems touch aspects of computer science or AI, (cognitive) psychology, and education. They are useful for understanding more about the processes of learning and educational interaction and can be used to test theories of cognitive psychologists, AI researchers, and educational theorists. Building an ITS is thereby not trivial but needs input from computer science, psychology, and education, especially regarding theories of learning (Gentner, 2003).

But what makes ITS so effective? VanLehn (2011) names three crucial aspects that might explain the superiority of human teaching and then argues how an ITS fulfills these three aspects as well: Firstly, human teachers give frequent feedback, allowing students to quickly find flaws in their reasoning and fix them. According to the Interaction Granularity Hypothesis, this is a crucial factor as the amount of reasoning necessary to understand one's misconceptions is low. The larger the grain size of the feedback to the user, the more reasoning is needed between the interactions. For example, if a tutor only provides the whole solution when an error occurs, a long line of reasoning is needed, so there is a large granularity. Tutoring systems also offer immediate feedback after each step, so there is a short line of reasoning as students can examine their first incorrect step and know right away where they went wrong. It thereby supports step-by-step problem-solving and problem selection. Hence, as the granularity of tutoring decreases, the effect sizes increase, thereby favoring both human tutoring and step-based ITS tutoring as they offer small grain sizes.[2]

Secondly, teachers use scaffolding or "a kind of guided prompting that pushes the student a little further along the same line of thinking" (VanLehn, 2011, p. 200). This technique often involves cooperative execution or coordination, thereby extending the student's reasoning proactively. Tutoring systems also scaffold students because if a step is accepted, the student knows that the tutoring system understands the step (coordination) and agrees with the reasoning (collaborative execution). If not, the system offers hints as another type of collaborative execution or support.

Thirdly, an intervening variable is suggested: tutors' behavior influences the students' behavior which affects students' learning. Ideally, a tutor should encourage students to engage in interactive or constructive behavior. As an ITS requires students to correctly solve the task before moving on, students solve their set of problems by doing most of the reasoning themselves. By providing enough scaffolding and feedback, they enable students to generate correct answers on their own, thereby encouraging constructive behavior.

Lastly, as an ITS only provides help when needed, they fall into Vygotsky's zone of proximal development where proper learning takes place (Myneni et al., 2013).

Thus, both human tutoring and ITS tutoring may be superior tutoring strategies as both are promising at scaffolding students and giving feedback to encourage students to engage in interactive as well as constructive behaviors.

2.2 Its Four Modules

An ITS is typically comprised of four components or modules which interact. Students engage with the ITS via an interface. Their actions are evaluated concerning the domain knowledge component (expert knowledge module). Based on this evaluation, a model of each student's knowledge status is maintained in the dynamic student model. Lastly, the pedagogical module delivers instructional actions based on the student model (Corbett et al., 1997). Typically, the knowledge and inference rules are defined by a designer, but the teaching sequence is decided by the system (the pedagogical module). I will now consider these modules in more detail.

Firstly, the (expert) knowledge module holds the knowledge of the tutored domain, the techniques of problem-solving, and the actual teaching material to

[2] The review of VanLehn (2011) also indicated there is no significant difference between a step-based and a sub-step-based ITS where each step is further divided, and additional feedback is given. I will hence implement a step-based ITS as they are easier to implement but just as efficient.

be conveyed to the student. It thereby explicitly represents the content knowledge so it can generate solutions to any of the problems the students face. It is sometimes considered the heart of any ITS as the module requires the development of a true cognitive model of the domain knowledge to solve exercises in the same way students do (Corbett et al., 1997).

Secondly, the student model module dynamically captures both the current knowledge as well as misconceptions of a student. It creates a qualitative representation of the student's cognitive and affective knowledge state, thereby providing each student's knowledge base as well as the conditions for adjusting feedback as it should respond effectively to the individual needs of the student. More precisely, three types of student models exist (Almasri et al., 2019). An overlay model compares the behavior of the student with the behavior of an expert and takes the difference as the knowledge the student lacks. It is a copy of the knowledge module in which each knowledge unit is tagged with an estimate of how well the student has learned it. Each student module has an overlay model, but some provide more: A perturbation model extends this approach to include a bug library, so a student is also judged regarding typical errors and misconceptions students make. Lastly, the learner-based model focuses on knowledge acquisition to prevent misconceptions: If a misconception happens, problem-solving rules explain the (erroneous) steps leading up to the misconception. I will use a learner-based model here.

Thirdly, the pedagogical (or tutoring) module realizes the principles of teaching by providing the infrastructure to select a teaching element according to the student model (and thereby the student's needs). It detects the knowledge deficiency, selects a matching tutoring strategy, checks for the learning process, generates constructive feedback, and guarantees the student's motivation. For example, this can be achieved (as is the case in my ITS) via a step-by-step approach where the program provides a sequence of actions for the student to follow, giving feedback or hints after each step (Corbett et al., 1997).

Lastly, for the system to interact with the user, an ITS comprises a user interface module (sometimes called a problem-solving environment). This should help solve the task by defining the problem-solving activities the student engages in by translating between the internal representation of the system and a human-readable interface language. It should be fitting by approximating real-world problem-solving environments and facilitating the learning process. One aspect is to ensure students can type in or click on everything they need to solve a task, for example by providing syntax shortcuts. Given the use case, it should hence be easy to type chemical symbols. Several types exist, such as graphic communication models (e.g., virtual reality approaches), social intelligence models (such as verbal analysis or speech cue recognition), component interfaces which process the student input and evaluate symbols, or natural language communication, such as different dialogue techniques (Almasri et al., 2019). I will use a component interface here as it is very suitable for working with chemical formulae. Furthermore, the focus of this work lies in the backend, as the interface provided is just a prototype and has not been subjected to quantitative usability studies (see Chap. 6).

2.3 Some Tutoring Systems in Practice

This chapter summarizes tutoring systems that are interesting to my use case. Accordingly, different kinds of tutoring systems and their typical areas of application are considered as well as systems relating to chemistry, and systems using analogies as a feedback mechanism.

There are a variety of ITS currently in use, some with quite interesting add-ons or teaching modules. In general, there are rule-based types (Corbett et al., 1997) and example-tracing tutors that rely on generalized examples of correct and incorrect solution paths. For example, one system represents the domain as a set of constraints specifying the correct patterns of punctuation and capitalization, and feedback is given on violated constraints. Another system uses Petri nets for the pedagogical model: The interaction between student and agent is controlled by an object Petri net and translated into a rule-based expert system. The transitions of the Petri net are controlled by logical conditions that refer to the student model and the firing of these transitions produces actions that update this student model (Almasri et al., 2019). Many systems provide feedback after each step by giving hints on incorrect steps already taken, or steps to be taken next if the answer to the previous step was correct. Some physical ITS (for example teaching thermodynamic cycles in closed systems) analyze students' submissions to give adequate feedback.

Other tutoring systems focus more on the pedagogic aspects. Notably, an early ITS from Anderson et al. (1985) is built on their ACT*-theory which is an attempt to identify the principal factors affecting human cognition and organize them into a complete cognitive theory. Another ITS aims to explain pulley systems by combining simulation and tutoring, customizes tutoring to address common misconceptions, and employs pedagogic strategies. They choose fundamental physics concepts students find hard to understand, and which are often not sufficiently addressed in schools (Myneni et al., 2013).[3]

Overall, there is a wide variety of existing systems where most tutoring systems seem to be concerned with learning a language (including programming languages)

[3] Note that this is very similar to my approach, except my domain is chemistry, not physics.

or software tools (such as Excel, Photoshop, etc.). Many ITS teaching mathematics or medical issues (such as diagnosing diabetes) exist as well as ITS on religious issues such as rules of the Holy Quran (Almasri et al., 2019). However, there are hardly any chemistry-related tutoring systems. Tallying with this, Mousavinasab et al. (2018) conducted a survey study on (English) papers evaluating tutoring systems published from 2007–2017. This survey confirms the lack of chemistry-related tutoring systems as chemistry is not one of the main educational fields targeted by existing systems although it is well suited as a field of application: "Although ITSs could facilitate reasoning in the learning process, these systems are rarely applied in experimental courses including problem-solving and decision-making" (Mousavinasab et al., 2018, p. 18).

In line with this finding, I can only report a few tutoring systems for chemistry here. One chemistry system called $ConnectedChemistry$ (Stieff, 2005) targets how different representations depict the same concept. They aim to improve students' knowledge of different representation schemes in chemistry and their structure. The $SVM : Chem$-system (Kozma and Russell, 2005) is used to teach how graphical and symbolic representations of chemistry concepts relate to one another. To make sense of connections between representations, $eChem$ (Wu et al., 2001) is designed. Perhaps the most recent ITS, Rau et al. (2015) started building an ITS called $ChemTutor$ for chemistry via a multi-method approach to help enhance connection-making between multiple graphical representations that are ubiquitous in STEM topics as these representations are critical for learning. Chemistry instructions especially heavily rely on the use of graphical representations. They found large and significant learning gains in performance, effectiveness, and efficiency measures assessing students' knowledge and reasoning. Lastly, $ChemLogic$ is a computer system based on logic programming (Prolog) to solve several chemical problems, such as stoichiometric calculations or completing equations (Paun, 2015). This is hence the only system aiming at chemical equations, but it is not a tutoring system for it lacks any feedback module but simply depicts the correct solution to a given query.

Furthermore, some tutoring system using analogies as a feedback method exist, yet, to my knowledge, none of them applies analogies to the field of chemistry. One system incorporates analogies and worked-out examples as teaching strategies in a computer science tutoring system (Harsley et al., 2016). Interestingly, students given worked-out examples outperformed both students given analogical worked-out examples and students given analogical explanations on a post- compared to a pre-test. However, the differences in learning gain were not significant. Using linear regression, the authors found a significant negative correlation between pre-test score and learning gain when both analogies and worked-out examples are used, suggesting that analogies are least beneficial to students with higher prior knowledge. Furthermore, the students given analogical explanations only performed worst. Hence, using worked-out examples (with analogies) can be considered superior to the mere use of analogies although the results are merely correlative. Nevertheless, I will use analogies with worked-out examples in my implementation.

As a second example, Lulis et al. (2004) built an ITS system for cardiovascular physiology using natural language dialogue. They used two types of analogy: The first, more straightforward idea, is to substitute a variable and lay out the behavior of the new, substituted variable. This is a rule-based system. The student should then—by analogy—understand that in the given scenario, all variables will behave in the same manner, so the behavior of the given variable has to match the behavior of the analog. Secondly, the authors also wanted to implement analogies outside their (physiological) domain. For example, they drew an analogy between a water-filled balloon and the distension of the right atrium. To do so, they used the Structure Mapping Engine (Forbus et al., 1997). In most of their examples, the mapping between target and base was successful in that students understood the analogy correctly, resulting in a clarification of the physiological problem (Lulis et al., 2004).

Lastly, another ITS, namely $SubKraki$, exists which uses the logic programming language Prolog to provide feedback via analogous examples (Zeller and Schmid, 2017). It focuses on learning written subtraction at primary school age. If a student makes an error while subtracting two numbers, the system presents an analogous example, that is, two numbers sharing the same features as the input to guide the students towards the correct solution. I will somewhat adopt this strategy but adapt the analogies and the error feedback to my use case.

Hence, there are some tutoring systems aiming to enhance students' understanding of certain areas of chemistry but no ITS is targeting the critical domain of reactions. The systems either engage in other fields or provide no tutoring. An approach to redox reaction is missing completely. Such a system is therefore much needed.

How to Identify Misconceptions in the Feedback Module of an ITS

3

There are different approaches to the design of the feedback module aiming to give good, immediate, helpful responses to the user of the tutoring system. In this chapter, I will briefly describe some commonly used types of error feedback that are part of many ITS, most notably error libraries, before analyzing the method of Algorithmic Debugging itself and its application to the field of tutoring systems. This will be laid out in more detail as my own approach builds on this technique though I use a simplified and adapted version which fits the requirements of my chemistry use case better (see Chap. 6).

3.1 Error Libraries

Firstly, an overlay model assumes that any discrepancy between student answer and expert knowledge is reducible to the student's lack of skill. It therefore only requires an adequate representation and operationalization of the necessary background knowledge, that is, a working expert module with factual knowledge units and procedural skills (Zinn, 2011). However, this underlying assumption is inadequate as studies suggest that student errors cannot solely be described in terms of absent (expert) knowledge of the task (J. Brown and Burton, 1978). Hence, an additional bug library containing typical and common student errors in a given domain is necessary as part of the feedback module of an ITS. The errors are recorded in a database, so that misconceptions are represented as a catalog of typical errors. Typically, the error type, frequent occurrence, and associated diagnosis are coded. An erroneous answer can then be reproduced by finding the same error (as a variant of the expert answer) in the error library (Anderson et al., 1985). Such a system supports specific, corrective feedback based on the correct solution to the error entry in the library.

M. Stümke, *Identifying and Overcoming Student's Misconceptions in Chemistry*, BestMasters, https://doi.org/10.1007/978-3-658-51032-9_3

However, the quality of the diagnosis is often inadequate, as either the set of faulty rules is incomplete or large amounts of data have to be recorded: Such a thorough cognitive diagnosis requires a high granularity of skill decomposition, and, thereby, an in-depth analysis of a large number of student errors and protocols. Put differently, as bug libraries can only detect the errors listed in the library, the approach is either highly incomplete, or computationally complex (Zinn, 2011).

Other feedback possibilities exist, for example, production rules systems. Such a system is executed to identify a rule that can reproduce a student's action and produces the feedback associated with the fitting rule. Constraint-based tutoring systems, on the other hand, test a problem state, the input of the student, against a set of constraints to test whether relevant aspects of the state are satisfied. An unsatisfied constraint is then considered a potential source of error and a corresponding error message is returned. Model tracing tutors only analyze the student's next step towards a solution and give immediate feedback after each step. This way, it is not necessary to reproduce their entire line of thinking, so this approach is less complex. On the other hand, such tutoring systems are complicated to build as they require both a precise cognitive task-analysis and an interpretable user interface design (Zinn, 2011). Moreover, as in all these approaches the steps are supervised rather tightly, students have fewer opportunities to explore different solution paths, as they are presented with corrective feedback right away.

3.2 Algorithmic Debugging

As all previously discussed approaches have serious disadvantages and shortcomings, the approach of applying Algorithmic Debugging to an ITS is presented here. It is in line with the feedback structures of both production rules and constraint-based approaches. But its basic approach is no longer step-based, so the computational complexity is low as neither dialogue interventions nor bug libraries are needed. Further, as it comprises a hierarchical representation of expertise, induces a human readable teacher-student dialogue, and sophisticated feedback can be given at any time, it outperforms the other approaches (Zinn, 2014). I hence planned on using Algorithmic Debugging in my ITS though some alterations turned out to be necessary during implementation (see Chap. 5).

The first subchapter lays out the basic idea of Algorithmic Debugging (Shapiro, 1982), before the second subchapter describes how to apply this technique to the feedback module of a tutoring system by pointing student's misconceptions to the execution trace of the expert module. Thirdly, the feedback module is further improved by adding analogical traces (Zeller, 2015). Lastly, the chapter concludes with remarks on logic programming Algorithmic Debugging builds upon. Using the methodology of logic programming, I will describe my approach to apply this framework to the use case of chemistry.

3.2.1 Algorithmic Debugging in Shapiro

Program debugging aims at bridging the gap between programmer and executable program. As a program is arbitrarily complex, however, one cannot generally anticipate any possible behavior of the program. Hence, debugging should be solved at a more abstract level, thereby giving algorithmic solutions to the questions of how to identify a bug in a program behaving incorrectly and how to fix such an identified bug. To do so, Shapiro (1982) implemented a semi-automatic debugging algorithm with a high level of abstraction and composed of two phases. It comprises an interactive, incremental diagnosis algorithm to identify a bug in a program using a dialogue between debugger and user, and a correction algorithm to fix a bug that has been identified. The algorithm receives a program to be debugged as input, traverses down the refutation tree (the data structure representing the execution of the program) and marks the corresponding nodes as correct or incorrect until a buggy node is found. A node is considered buggy if it leads to an incorrect result, yet all of its children (if any exist) are correct (Insa and Silva, 2018).

More specifically, algorithmic debugging (AD) defines a systematic approach to identify bugs or misconceptions in programs based on a dialogue between the programmer and the system. The program is represented (structurally) as a tree with the goal clause as the root. Using a top-down approach,[1] the tree is then traversed

[1] This is not the only approach offered by Shapiro (1982). He also considers other variants or extensions, including a bottom-up variant, minimizing the number of queries to the oracle, and procedures terminating with incorrect output or non-terminating procedures. But as none of these variants are of importance for my implementation of an ITS, I will only explain the top-down variant here.
Furthermore, some thoughts on reducing complexity complement the work of Shapiro (1982): For example, his single-step navigation technique through the tree to find the bug is improved by pruning the debugging tree using a divide-and-conquer strategy, thereby putting a lower bound on the number of queries and reducing the size of the tree by nearly a half after each answer, so bugs can be solved in polynomial time only. Also, the more the oracle is mechanized, the fewer user input is necessary. This can be done, for example, by storing answers between and within sessions, so the user is not asked the same question twice. Nevertheless, time and memory store are still an issue, the granularity of errors is too large, and the oracle may ask overly complicated questions to the user (Caballero et al., 2017). However, I will not expand on the shortcomings or the improvements here.

from the goal clause to its roots. This debugging tree—the internal representation of the logic behind a (faulty) computation—is obtained using a meta-interpreter as a reflection possibility of the logic programming language Prolog. At each step, the programmer takes on the role of the oracle to answer if the goal holds or not. If both programmer and system agree the goal holds, the algorithm continues onto the next goal on the goal stack. In case of a disagreement, the goal is traversed down until an irreducible disagreement between programmer and system is discovered as a child node of the goal. This bug of the system can be localized as the program clause currently traversed, so the irreducible clause is localized as the source of the buggy behavior (Shapiro, 1982). The method thereby extends a meta-interpreter for logic programs: While the program is meta-interpreted, oracle questions are generated, and the program is either silently executed or investigated further, if a disagreement was found (Zinn, 2011). To do so, a debugging algorithm comprising an interactive, incremental diagnosis algorithm to identify a bug in a program, and a correction algorithm to fix a bug that has been identified, are implemented.

Firstly, the diagnosis algorithm is interactive in that it queries the programmer to check if the results of the program are correct or not and aims to detect errors in a program that terminates with an incorrect output.[2] It iterates stepwise over the procedure calls of the program and returns the first program to return an incorrect output. The output values are checked against the "*yes*" or "*no*" answers of an oracle (either given automatically by the algorithm or interactively by the user). Hence, if the oracle returns "*no*", the corresponding procedure is detected as incorrect. The overall diagnosis system then starts with a loop where the algorithm iterates over all goals, reads and solves them. In the event of an unexpected output value, i.e. an incorrect result, the algorithm runs through the program step by step (iterating down the tree) and iteratively asks the oracle whether the respective step is correct. The processing happens from left to right and top-down. In this way, the tree is traversed until an error is found. Here, the children are again considered individually until an error can be localized recursively at the lowest level of the tree (Zeller, 2015).

The second component of algorithmic debugging is the bug-correction algorithm, which I will only quickly lay out here as, again, it is not immediately relevant to my implementation. The algorithm aims to find a close variant to the faulty clause at hand that is correct. In other words, if the clause fails on a goal known to be true, the algorithm searches the equivalence class of the current clause and proposes

[2] To detect an incorrect output, a formal definition of correctness is needed. This definition states that a procedure p is correct in an interpretation M if any output covered by p with respect to M actually is in M. put differently, if a procedure is incorrect, there is an output-trace in M of an output not in M. This trace then provides a counterexample to the correctness of the procedure (Shapiro, 1982).

a correct clause within the class as a refinement, so the goal evaluates to true. If a clause considers an incorrect goal to be true, however, a clause in the program is considered false. If no substituting clause can be found, the user is asked for a possible solution to the bug. The correction algorithm is thereby semi-automatic, as the class of bugs the algorithm identifies automatically is restricted, and more complex bugs are left to be corrected by the oracle. Lastly, the debugger either returns this buggy node to the user or $\bot$ if no such node exists (Caballero et al., 2017).

Put together, AD is composed of a diagnosis algorithm to identify and locate the errors, an interactive shell needed to ask queries to the user and store their answers, and the bug-correction algorithm intended to correct the errors. It is a search strategy aiming to automatically find and correct programming errors (bugs) in a program whose input is an execution tree and whose output is either a buggy node or a message indicating that no bug exists to automatically find and correct programming errors (bugs) in a program. Although this original idea influenced the development of new debugging systems in logic programming as well as other programming paradigms, such as declarative programming or object-oriented programming (Caballero et al., 2017), the technique is barely put in practice as errors are not precisely located within the code. Usually, algorithmic debuggers do not point to a line of code as buggy, but rather to a whole procedure. Hence, the procedure needs some adapting. One such adaptation approach is presented next.

3.2.2 Algorithmic Debugging Applied to an ITS

Program debugging is a crucial part of the domain expertise (expert module) of an ITS as it serves as (part of) the expert module communicating with the student module to evaluate their answers. Ideally, student programs are automatically debugged, so that the bug is identified and exemplary solutions specifying correct answers and ideas to overcome the bug are given. This way, intelligent hint generation assists students to understand the task at hand. To do so, a methodology and implementation for the diagnosis and correction of nontrivial recursive programs—including algorithm recognition, bug detection, and bug correction—is needed. This is complex as there is a tremendous variability in student's solutions to nontrivial tasks, for example by omitting necessary procedures or adding superfluous steps (Murray, 1987).

A possible approach is an ITS based on logic programming, so the expert knowledge module can be represented as a (logic) Prolog program and be based upon the encoding of expert problem-solving. This can be compared with the solving steps of a learner to diagnose their errors. The idea is that any deviation between the correct step predicted by the expert module and the student's solution reveals an error in the learner's reasoning. Zinn (2014) extends the algorithmic debugging approach and adapts it according to this line of reasoning: AD is used to incrementally meta-interpret (see Chap. 4) the expert module by comparing its behavior with the behavior of a student using a mechanized oracle. The program location of any deviation is then captured to know where the student's error occurred.

To adapt Shapiro's (1982) approach to an ITS, it is not for the programmer to specify how the presumably incorrect program should behave, but rather for the correct program (the expert module) to locate the errors made by the student. The algorithm is hence fulfilled differently: The expert program serves as the buggy program, while the oracle is a student's answer. This is to say, the answer to the oracle queries is derived from the student's answers to the problem at hand. If an irreducible disagreement occurs between the expert module and the automatic oracle, it is thereby located within a clause of the expert module. As the student misconceived this step in their line of reasoning, the correct approach to solving the underlying logic of the clause should be returned as feedback to correct the error. To do so, the algorithm `algorithmic_debugging/1` traverses a given program top-down and is called for each goal of that program. Firstly, the algorithm checks if the goal should be discussed. For example, if it is a Prolog built-in predicate, a mere technical necessity, or a goal set by the user, it should not be discussed as it cannot be the source of the error. If, however, the goal may result from a misconception, the algorithm looks for an agreement or disagreement between the expert module and the student's answer (the oracle). In case of an agreement, the goal is considered processed, and the next goal on the stack is recursively called. In case of a disagreement, the goal structure is traversed down until an essential clause marked as *irreducible* is found. This predicate always evaluates as *true*. The original algorithm from Zinn (2011) terminates here by returning the clause in question and formatting it into a human-readable statement using the `get_diagnosis/2`-predicate. The system allows to track multiple errors, which are repaired one-by-one. If, for example, students correct their first error, but make a second one during this correction, or their correction is incomplete, the predicate can be called again and starts traversing the new tree to detect the second error. This process continues until no more errors are made by the student.

However, simply returning this disagreement in natural language feedback may be shortsighted as feedback should always be embedded into the context (Zinn, 2014). For example, it is often the case that not the most recent step is faulty, but the result of a prior, faulty step. Hence, an algorithmic debugging approach terminating at the first irreducible disagreement and generating only one feedback comment is not sufficient.[3]

Zinn (2014) implemented comments into his expert model to annotate the faulty steps. In other word, the source code is annotated in such a way, that the output links to the misconception, but the comment is ignored by the Prolog compiler and interpreter, so it does not affect the execution of the expert module. If several comments are added to one clause, they are ordered from low to high specificity. Hence, the first comment in a clause provides less (specific) information about the error than the second comment, etc. The execution trace outputs the comments. Consequently, after detecting a disagreement, the entire program is traversed by iterating over the goal structure or the atomic goal and checking if a disagreement between the expert module and the oracle occurs. Every node of the execution tree is thereby marked as either agreeing or disagreeing with the student's behavior. The execution tree starting from the goal is explored until a disagreement marked as *irreducible*, an atomic disagreement, is found. This way, all program clauses expert module and student agreed or disagreed upon are obtained and the hierarchical context is included. This list of agreements and disagreements is passed onto the feedback engine as the hierarchical embedding if help is requested by the user. This means that the error in one of the predicates can be localized and labelled at the same time, which is a decisive advantage of the approach compared to error libraries. At the same time, this approach is also limited in terms of the number of errors that can be diagnosed, as only errors marked as *irreducible* in a predicate can be detected. Although the approach contains information about previous steps in solving this task, the approach to previous tasks is not considered, which poses a second limitation (Zeller, 2015).

Lastly, based on the detected faulty step, the feedback engine generates hints based on the hierarchical context of the program clause in question as well as the learner's past performance (Zinn, 2014). If the clause is a parent of child nodes, all annotations of both parent and children are returned as feedback. If, on the other hand, the clause is a child node, only the least specific feedback of that clause is returned. If help is requested again, the second least specific feedback is returned, etc., so a different disagreement structure is returned until the task at hand is solved. In other works, only if help is requested, the algorithmic debugging starts anew, considering everything that happened before this request (for example, that the least

[3] More precisely, the use-case here is written, multi-column subtraction. Hence, each step of a student aiming to solve a written subtraction task is compared to the correct solution represented by the expert module.

specific feedback has already been given to the student, so the student needs more specific help).

One important difference to the original approach is that the process of "asking the oracle" is mechanized as it seems highly ineffective to ask the students if their solution agrees with another solution after each step: The implemented oracle therefore takes the full answer given by the student, and automatically extracts those parts required to give the answer. The predicate `oracle/2` is called after each step executed by the expert model and returns whether the student agrees or disagrees with this step of the expert module ("*yes*" or "*no*"). The currently processed column is marked and compared to the length of the answer. If they are equal, the answer is compared to the expert module to detect deviations. If not, the algorithm recursively continues comparing the next column to the answer until a match is found.

Put together, algorithmic debugging is similar to model tracing in that help is provided with each step if a disagreement was found. The great advantage lies in its simplicity due to including a meta-programming perspective (Zinn, 2011). It also does not require a representation of typical errors, but the feedback may be less effective because of it. Additionally, step-based approaches within in an ITS have outperformed other approaches (VanLehn, 2011), so it would be useful to adapt AD to allow sub-steps. All in all, algorithmic debugging alone is thereby not sufficient to fully grasp all misconceptions, so it might be fruitful to combine it with other (logic) techniques to acquire a fully sound methodology of cognitive diagnosis. One approach is to combine it with analogical problem-solving, as Zeller (2015) did in her master's thesis. In the next subchapter, I will therefore lay out how the analogy-based feedback approach works before describing how I linked these strings together in my own approach.

3.2.3 Improving the Feedback by Adding Analogical Traces

Previous experience with analogous tasks is an important source of information for solving present tasks. People also consider analogies especially if they fail at a task (Gick and Holyoak, 1980). In this respect, presenting an analogous example seems a promising way to give students feedback (see Chap. 4). Again, one could use the principle of a library, i.e. create a library of analogous tasks containing a suitable counterpart for all diagnosed errors. However, the problem with libraries is still the same: If only a few tasks are included, the approach is inflexible and if many analogs are included, it is complex. For this reason, an attempt is made here to automatically choose a fitting analogous example to the task and error at hand. Since good analogies are structurally similar and an algorithm describes a sequence or procedure, an analogous example should follow the sequence of the algorithm for solving chemical formulas. This way, a structurally similar example is ensured, and the chemical substances of the analogous formula can vary, thereby not affecting the structure.

3.3 Adapting the Use Case to Prolog

As the backend—and therewith the expert and feedback module—is implemented in Prolog, I will briefly point out some important concepts and features I used in my implementation.

Prolog is a declarative logic programming language based on a resolution calculus (Warren, 2023). It is designed to make true statements about a world and to answer questions (queries) about this world. Its syntax consists of literals, a single predicate that is either negated (negative literal) or not (positive literal), commas meaning $\wedge$ (conjunction), semicolon meaning $\vee$ (disjunction), the token :-, and a full stop as the end of each clause. An anonymous variable that appears only once in a clause is denoted by a single underline "_". The cut function "!" is used to reduce the search space of the computation by pruning the search tree, cutting off the currently traversed subtree if following along that path is known to be fruitless. Lastly, lists in Prolog are defined recursively based on the empty list denoted as []. The notation [Head | Tail] represents a non-empty list with its first element Head and a list Tail containing the rest of the elements.

Furthermore, predicates such as constants are denoted in lowercase, and variables in uppercase. A clause is then a disjunction of positive and negative literals. More precisely, Prolog uses Horn clauses defined as a set of disjunctively connected literals with at most one positive literal (Sterling and Shapiro, 1994). The main constructs, consisting of these Horn clauses, are thereby queries, facts, and rules. They make up the knowledge base representing the world. More precisely, rules are logical statements (Horn clauses) of the form

$$A \leftarrow B_1, B_2, \ldots, B_n \quad \text{with} \quad n \geq 0$$

with A (an atomic formula) as the head of the rule and $B_1, B_2, \ldots, B_n$ with $n \geq 0$ as the body. The token :- is used to denote logical implication and can be considered a reversed implication arrow ($\leftarrow$.). It hence reads "A holds if B_1 to B_n hold". Given $B_1, B_2, \ldots, B_n$ we can thereby deduce A conditionally via a finite number of applications of modus ponens (Sterling and Shapiro, 1994). A fact such as $father(abraham, isaac)$ is simply a rule with an empty body. There are no antecedent conditions, so it is unconditionally true in the given knowledge base.

Given such rules and facts in Prolog, queries about them can be answered using an inference rule called resolution. Queries can be either facts, to check for their truth or falsity, or they can be composite expressions to retrieve terms that make those expressions true. Resolution is a top-down and left-to-right evaluation, as it starts at the top with a query and computes down to the facts (Warren, 2023). It uses unification, a substitution between two terms to make the term identical. The most general unifier (mgu) of two terms is a) a unifier and b) the associated common instance of the terms is the most general one (Sterling and Shapiro, 1994). Prolog searches for precisely this mgu in its resolution calculus: At each step, a goal in the query and a clause in the knowledge base are chosen such that the head of the clause unifies with the goal using the mgu. When it fails to find such a clause, it uses depth-first backtracking to the last choice point, trying to find a subsequent clause to match the query. If such a match can be found, the program returns $true$ and if no unification can be found, $false$ is returned.

Debuggers are also based on Horn clauses and consider it a wrong answer if the user obtains an unexpected, computed answer for some initial goal. The computations correspond to predicate calls: The result of a computation is a substitution of logic variables with terms which may lead to obtaining the wrong answer. Each node is associated with a predicate clause which solves the predicate call producing the result. A computation then depends on the computation of the predicate calls associated with the atoms occurring in the body of the predicate clause (Caballero et al., 2017). Furthermore, a debugging tree in logic programming is a proof tree, thereby formally proving the correctness of the technique. An intended interpretation is a set of ground atoms (atoms containing no variables). If the program is correct, then the set of answers for a given predicate call exactly covers the set of instances that are true in the interpretation. On the other hand, a wrong answer occurs if there is some index and some substitution possible that are not part of the interpretation. In this case, the AD ends by pointing to the incorrect predicate clause. An error message typically indicates the predicate, the particular clause, and, ideally, the instance associated with the buggy node to locate the error precisely.

Moreover, one can easily manipulate its own programs as a program is just a data structure in Prolog language. Put differently, a Prolog feature is the equivalence of programs and data as both are represented as logical terms. Thereby, programs need to be treated as data, and data must be transformed into programs (Sterling and Shapiro, 1994). Such manipulation is known as meta-interpretation and is somewhat comparable to passing procedure names as parameters in object-oriented programming. It can be used to add logging subgoals to provide a trace, a proof tree, or modify it to perform different kinds of evaluation. Also, any debugging is meta-programming as it manipulates, analyzes, and simulates other programs and thereby part of the ITS by Zinn (2014). But how does it work? A meta-interpreter is a program that takes another program with a query as input and evaluates that query with respect to the program (Warren, 2023). For example, the meta-logical type predicate `var/1` tests for any given term if it is an uninstantiated variable or not and ==/ checks if two variables are identical. Such meta-logical predicates thereby consider variables rather than terms and treat other programs as the data they analyze.

Lastly, Prolog files can be organized via modularization. A file can be declared a module at the top of the program. Modules essentially allow you to hide predicate definitions, so some predicates within a file can be made public by making them part of the module. All other predicates are then private, so they are only callable from within the module itself, not from other parts of the program. The `use_module/1`-predicate allows you to load these modules into another file, so the necessary code parts can be used there as well.

3.4 Chapter Summary

Misconceptions can be represented explicitly in error libraries, but this is often unwise as the set of faulty rules tends to be incomplete. AD allows for more flexibility. Here, every possible error in arbitrary problems can be identified based on the expert production rules, so the system is not restricted to a predefined set of problem. Given any input/output examples (for example, chosen by the student), a program is automatically tested and the program part responsible for an error is identified. Zinn (2014) applied this to an ITS. It allows to compare the rules represented in the expert knowledge module with the student model module. So, if the two solutions differ, the difference can be described by the failing rule in the expert module which is wrongly represented in the student module.

The tutoring module then offers a strategy to help with the underlying misconception, so the student receives immediate feedback. In my tutoring system, Prolog is used to give this feedback by providing a structurally analogous example along with an explanation to help students understand the errors they made. I will now explain this analogical feedback in more detail.

Providing Feedback with Worked-Out Analogical Examples

4

This chapter elaborates the feedback module of my ITS further. As the feedback module uses analogies, the first subchapter explains what an analogy is while the second subchapter explains how analogy relates to learning—and, thereby, why it makes sense to use analogical examples for the pedagogical module of an ITS. The third subchapter relates analogies to worked-out examples to further enhance the learning effect of analogies.

4.1 Defining an Analogy

> "Longbottom, if brains were gold, you'd be poorer than Weasley, and that's saying something" (Rowling, 1997, S. p. 240).

The example above constitutes an analogy as the structural layout allows to draw a connection between the two objects it refers to. More precisely, Draco Malfoy builds an analogy between the poverty of Ron Weasley and the foolishness of Neville Longbottom. This is a rather straightforward, every day and common-sense use of an analogy. Comparing two objects or ideas like this is common practice in our use of words as well as routines and help us to communicate easily (Pritchard, 2018).

Recall Einstein's quote that a telegraph is just like a very long cat. It shows how an analogy can also be used to explain a new and complicated technology, to understand abstract and higher-order relations between objects. Interestingly, the message is seemingly effortless for humans to understand although the task at hand (to understand the semantic similarity between the objects) is a complex one

M. Stümke, *Identifying and Overcoming Student's Misconceptions in Chemistry*, BestMasters, https://doi.org/10.1007/978-3-658-51032-9_4

(Gentner, 2003). In fact, reasoning by analogy can be one of the most complex and highly adaptive cognitive processes in abstract thinking (Flemming et al., 2013). Put together, analogies are considered part of both common-sense and abstract reasoning which makes them a powerful cognitive mechanism to make inferences and learn new abstractions.[1] They have even been labelled a core mechanism of human cognition (Gentner and Smith, 2013) as human thought is specially characterized by the ability to engage in analogous reasoning (Pritchard, 2018).

But what defines this special kind of reasoning? An analogy is the comparison of two structured representations where the knowledge about a base domain is mapped to a less familiar (or more abstract) target domain (Gentner, 1983). To solve an analogical problem, one must determine the relational concept instantiated among one set of stimuli and subsequently search for that same structural or functional relation among sets of novel and unfamiliar exemplars (Flemming et al., 2013). It can be formally characterized as "A is to B as C is to D":

$$A : B :: C : D \tag{4.1}$$

This implies an arithmetic proportion as 4.1 states the equality of two differences:

$$A - B = C - D \tag{4.2}$$

Solving an analogy via analogical reasoning is a kind of inference (Bounhas and Prade, 2023): Any analogical equation asks to find a value X for a given triplet (A, B, C) such that the following equation holds.

$$A : B :: C : X = \top \tag{4.3}$$

Analogical reasoning is thereby considered a search for a solution in a space defined by prior domain knowledge, so analogies can be solved via exhaustive search. In doing so, they entail background knowledge (cf. of rules), deductive or symbolic reasoning to apply these rules correctly, and pattern recognition of the input samples. For example, to understand the analogy "3 is to 6 as 2 is to 4" described above, we need background knowledge about numbers, deduct the arithmetic properties, and recognize the shape of the numbers (Gentner, 1983). We do not, however, need

[1] There is, in fact, long-standing psychological evidence that reasoning by analogy helps humans to solve such complex problems (Gick and Holyoak, 1980). The evidence suggests that even young children can identify analogous relations (Flemming et al., 2013, Pritchard, 2018) and that they are especially helpful for students lacking prior knowledge (Forbus et al., 1997).

to compare similar features. This is what distinguishes analogical reasoning from similarity comparisons: Its core lies not in similar features, but in an alignment of structure.[2] Consider again the arithmetic analogy above. We do not care how many features the number 3 has in common with the 6 or the 2 with the 4 but only the relationship "twice as great as" is important here (Gentner, 1983).[3] We understand this abstract relation as we understand that two different concepts can play the same role and can align in their structure. Hence, analogy is about a common pattern of relationships among their constituent elements, not about their (perceptual) similarity. Base and target problems thereby have a structural similarity or an isomorphic structure, so there exists a one-to-one correspondence (a mapping) between them that preserves a binary relationship or structure between the elements of the sets. This relationship often depends on higher-order relations (relations between relations), such as causal, mathematical, or functional relations.

Aligning two problems structurally lies at the core of analogical thinking and led to Gentner's (1983) famous "structure-mapping" theory. According to the structure-mapping engine, the mapping is guided by the systematicity principle: According to this principle, if several comparable mappings are possible, the richest and deepest relational match is preferred (Gentner, 2003). An alignment associated with higher-order relations is thereby preferred over aligning single, isolated structures. Wiese et al. (2008) conducted an experiment supporting this claim: The subjects inferred knowledge about all concepts that could be learned, not just about those concepts they needed right away. They greedily mapped the largest possible structure from base to target and transferred as much knowledge as possible, thereby enlarging their possibilities to learn (see Chap. 5). Consequently, not only those parts of the

---

[2] More precisely, an overlap in both attributes and relationships is considered literal similarity whereas an overlap in only the relationships is considered an analogy. If only the attributes but not the relations overlap, Gentner (1983) speaks of mere appearance while a comparison sharing neither attributes nor relations is called an anomaly. Useful analogies thereby have larger numbers of shared relations but few common attributes.

[3] Note that the example taken here from Gentner (1983) is not as straightforward as it seems and thereby underlines the difficulty of coming up with a good analogy. Consider the three different formalizations of an analogy depicted above. Each is correct, of course, yet applying these formalizations yields more than one result: "3 is to 6 as 2 is to 4" is correct on all three formalizations but the relation of "twice as great as" is only unambiguously clear once you have C and D defined as 2 and 4. However, if we take (4.3) with "$(A = 3) : (B = 6) :: (C = 2) : (X = ?)$", it is perfectly plausible to assume a simple relation of subtraction so X could be 5 as well (and there may be other solutions). This often happens when creating a good analogy: The relation one aims to draw is plausible and can be deduced from the example, but further rules or connections are overlooked, so the analogy is not as unequivocal as hoped for, so finding suitable analogies is even more crucial. I will shortly draw on this finding in the discussion (see Chap. 6).

base domain relevant to solving the target problem at hand are transferred, but all inferences about the target domain are made which can be derived (directly or via higher-order relations) from the base domain (Wiese et al., 2008).

4.2 Analogical Reasoning and Learning

In general, when designing a learning intervention, it is always important to keep in mind theories of how knowledge is acquired, of how learning takes place (Rau et al., 2015). One such theory of learning, the one I chose for my ITS, is the learning theory of analogies. An ITS should foster learning, including learning declarative facts, associating facts with problem-solving goal structures, transferring knowledge to other environments, and acquiring meta-cognitive learning skills (Corbett et al., 1997; see Chap. 3).

It makes sense to choose analogies as a feedback method as they ensure these aspects of learning: Firstly, empirical findings indicate that worked-out examples which are analogous to an erroneously solved problem provide helpful feedback. Problem-solving is better facilitated by analogous examples than by written explanations (S. Brown and Salter, 2010). Students are also more motivated to engage with a problem if (analogous) examples are given. They can shorten practice time, reduce errors, and thereby enhance learning (Zeller and Schmid, 2017). Complex analogies can be used in learning via adapting analogs when deriving solutions to novel problems.

However, the mapping just explained does not suffice to explain how analogies foster learning as several subprocesses take place here. According to Gentner's (1983) structure-mapping theory, an analog (target) to a given situation (base) is found during retrieval (the first phase) before the mapping takes place. Based on this mapping, analogical inferences are made in the next step. This is called adaptation, and it creates new knowledge as the inferences are evaluated and fitted to the target's requirements (Novick and Holyoak, 1991). Next, abstraction takes place where the entities of the base and target problem are replaced with variables to encode schemata. This fifth component, the formation of an abstract schema for the problem class which facilitates generalization and subsequent problem solving, is where the actual learning takes place:[4] Drawing an analogy between two examples leads to a structural alignment—a set of correspondences between the elements of the two analogs in which their shared relational structure is highlighted—and to

[4] Lastly, Gentner (1983) states that a re-representation may take place which alters the content of base or target to improve the match.

the development of more general problem-solving schemata. This way, whatever is understood about one example can serve to shed light on the other example: Analogical encoding promotes the abstraction of schemata, which in turn promotes recall, transfer—and learning (Gentner, 2003).

Furthermore, as mentioned above, providing base and target analogies allows for structure generalization, so it allows one to not only solve the problem at hand but also build robust general solution routines. Such schemata allow the transfer of solution steps to other problems via anti-unification. Subterms, which are the same across different problems, are preserved whereas differing subterms are generalized, e.g., by introducing a new variable. Anti-unification therefore takes two terms and generates a common, generalized structure where differences are abstracted from and replaced by variables (Sterling and Shapiro, 1994). As a result, the least general statement—that is, the one with the fewest variables—is used to combine base and target problems while preserving the shared structure (Gentner and Smith, 2013).[5] Thereby, a generalized structure is obtained which is then filled in with another formula. Schema induction is an important consequence of analogical transfer as it not only includes the induction of more abstract and flexible knowledge about a class of problems, but schema quality also mediates the quality of analogical problem-solving and, therefore, of learning. Analogical transfer leads to schema induction as schema quality was positively related to the strength of analogical transfer for the target problem (Novick and Holyoak, 1991). This last aspect is of special importance as chemistry research points out the need to not only learn chemical formulae by heart but also understand the underlying mechanisms by building general solution routines (see Chap. 5).

4.3 Further Advantages of Analogical Learning

Using analogies fosters learning in that not only does the performance improve (as already shown), but also the competence. Competence is defined as the ability to do something well or effectively. That is, the structurally analogous transfer enables one to solve different tasks and to acquire general schemata, thereby allowing one to answer not just the example at hand, but related problems as well (Novick and Holyoak, 1991). For example, analogies play an important role in the development

[5] Anti-unification is the counterpart of the more well-known process of unification, which is a syntactical process which takes two or more terms to compute a new, single term. Variables are substituted with facts, so that the statement is specified or instantiated. For example, given $mother(x, Paul)$ and $mother(Mary, y)$ and instantiating the variables of x and y to obtain the term $mother(Mary, Paul)$ is a unification (Warren, 2023).

of intercultural competencies (Abe, 2019). They can be used to ease the transition from an experience in one's own culture to new, unfamiliar interactions with other cultures.

This improvement in competence points to the relational aspect of analogical ability. Relational learning is especially important to the development of cognition as children often learn by comparison or by distinguishing one thing from another. For example, they learn that an animal is not a dog, but a wolf because it is slightly larger and wilder. Such a comparison furthers the acquisition of knowledge, mainly by enabling schema abstraction and by promoting attention to relevant differences between the two concepts (Gentner, 2003).[6] More precisely, the results of several studies indicate that analogical encoding via making comparisons is a key driver of immediate learning by fostering representation in terms of a more general and complete relational schema or a higher schema quality, which is then more likely to be retrieved in new situations, and hence improves learning. These comparisons also improve learning as—common to all analogical approaches—they promote attention toward common principles or schemas. By focusing attention on structural similarities, greater results are produced. Lastly, this better learning in turn leads to a superior transfer (Gentner et al., 2003).

Moreover, it can be assumed that analogies foster self-efficacy. This concept refers to a person's degree of confidence in their capability to successfully perform a particular task or accomplish a specific goal (Bandura, 1977). They are one of the most powerful motivational predictors of how well a person performs at a task and the extent to which they engage in challenging tasks. As self-efficacy is therefore important for any self-regulated learning, it plays a crucial role in science learning (Crippen and Earl, 2007). Analogies are likely to have a positive effect on self-efficacy as they strengthen one's competence (see above). Thereby, students realize they can solve tasks they were unable to deal with before. If they attribute this change to their skills and abilities, they may well increase their self-efficacy scores. However, empirical studies investigating such a relationship are mostly lacking. Applying analogies to an ITS may therefore also be considered an appeal for empirical studies on self-efficacy and analogies. Some initial studies exist already, indicating an effect. For example, the use of analogy resulted in higher self-efficacy scores compared to both explicit and implicit learning in the acquisition of a new motor skill (Lola and Tzetzis, 2020).

---

[6] Next to analogies, Gentner (2003) argues for the importance of language, especially relational concepts of language, to learn by comparison. According to this theory, the process of extracting relations via analogy and then preserving them via language uniquely fosters learning and creates the structured symbolic representations essential for higher-order cognition.

4.4 Analogies and Worked-Out Examples

But what type of examples are needed to fully utilize said advantages of an analogy? In general, examples are an important tool in the early phases of cognitive skill acquisition. Simple, familiar examples can serve as models or analogies to deal with unknown situations. This method, however, is highly limited as people, especially novices, often fail to recognize or access relevant examples that would be highly useful if retrieved (inert knowledge problem; Gentner et al., 2003). A superior type of example is therefore needed.

A worked-out example (WOE) is a problem-solution-representation constructed in such a way that the student understands how the solution was established without providing a script or algorithm, but rather by providing and explaining the necessary steps. Several studies showed that worked-out examples aid learning and tutoring, especially in well-structured domains (Harsley et al., 2016). For example, a snippet of computer code can serve as a worked-out example to understand a programming task. They have shown to be successful in STEM-related topics for both experienced and inexperienced students as they lessen the workload and help to focus on problem-solving techniques which are thereby learned more quickly (Crippen and Earl, 2007). For example, results of a quasi-experimental study in chemistry indicate that combining worked-out examples with a self-explanation prompt (an iterative personal dialogue students can have with themselves while solving a problem) produces an improvement in the participant's performance, problem-solving skills, and self-efficacy (Crippen and Earl, 2007).

Furthermore, modular worked-out examples have repeatedly outperformed the traditional molar ones. A molar WOE treats problem categories as basic units that cannot be broken down further and can thereby become overwhelming. For example, in computer science one may be presented with a "recipe-like" formula including all the necessary solution steps to solve a task. One might memorize this recipe rather than consider and understand the procedure behind it. Hence, a modular WOE focusing on the role of individual structural problem features and solution steps is more fruitful (Gerjets et al., 2006). One such solution step is part of the overall solution procedure but can be understood in isolation, is transferable to new problems, and helps to understand the specific relations. A modular WOE also decreases the intrinsic cognitive load by breaking down complex solutions into smaller meaningful solution elements, so only a couple of elements, not the whole solution, have to be memorized (Gerjets et al., 2006). Hence, I use modular worked-out examples in the ITS.

Lastly, as mentioned before, the most important prerequisite for problem-solving is having abstract problem-type schemata (Gick and Holyoak, 1980). If such schemata exist and a problem has been identified as belonging to the category of the schema because of its features, the schema is instantiated with the information specific to the problem at hand. Worked-out examples, especially combined with analogies, can be a useful instructional device in constructing such schemata (Gerjets et al., 2006).

4.5 Using Analogies in the Feedback Module of the ITS

Overall, in the case of an ITS, a useful analogy captures the characteristics responsible for the diagnosed error and provides a structurally isomorph example so the student can transfer the solution steps from the example onto the current (target) problem. This is what the feedback module of my ITS aims to do. It further draws on scientific findings suggesting how best to use analogies when teaching chemistry. Analogies are useful, according to an interview analysis (Orgill and Bodner, 2004), if the target concepts are difficult or challenging, if the target concepts cannot be visualized, and if new target concepts are introduced. Moreover, an unsuitable analogy is compley or lengthy. On the other hand, good analogies are simple, easy to remember, have a familiar analog concept, and the relationship between analog and target concept is explained.[7]

Drawing on these research findings, it is plausible to assume analogies are useful as a supplementary method in teaching redox reactions as it is a difficult concept many students struggle with (see Chap. 5), it cannot be properly visualized in that you cannot see the actual processes but need models to approximate them, and the reactions need to be understood rather than memorized. Lastly, combining analogies with worked-out examples is fruitful in capturing the potential of both. An analogical worked-out example (AWOE) is thereby, structurally, a WOE, but the content is based on an analogy. Therefore, in my tutoring system, in each step when solving an equation, such an AWOE is provided.

[7] Put differently, analogies are inherently limited as they are not considered to be useful if a concept is simple to understand or has to be memorized. They may also hinder learning if the target is already understood and may be misleading if the mapping was not successful. Brown and Salter (2010) point out that many commonly used analogies in teaching may evoke a wrong understanding if misapplied as, for example, a heart does not resemble a pump in many ways. This again points to the necessity of clarifying the shared relations between base and target to ensure an understanding of the underlying structure.

"Chemieleon"—A Prototypical ITS to Teach Chemical Equations 5

This chapter comprises the approach to building and coding the ITS. The first subchapter introduces the use case of redox reaction and explains why an ITS could help overcome typical problems here. The next subchapter lays out the architecture of the system as well as the concrete implementation of each module. A special focus will be laid on the background knowledge module using logic programming to learn the rules and facts needed to solve equations as well as on the feedback module generating analogous examples.

5.1 Introducing the Use-Case Domain of Chemical Formulae

A scientific model is an approximating, pragmatic thinking tool that is of special importance to chemistry as the domain involves many complex and abstract concepts a model aims to simplify. Using models appropriately helps to visualize these abstract concepts (Coll, 2004). However, learning the proper use of models is a complex task similar to that of learning a language. Chemical equations are such a scientific model and thereby offer great potential to help understand the underlying abstract concepts. However, many students struggle to grasp chemical reactions, especially redox reactions (Hansen, 2022).

Accordingly, redox reactions are a very important topic in chemistry, yet very difficult to teach: Various studies reveal that students have difficulties in interpreting and using chemical symbols, formulas, and equations correctly. It is often assumed

Supplementary Information The online version contains supplementary material available at https://doi.org/10.1007/978-3-658-51032-9_5.

M. Stümke, *Identifying and Overcoming Student's Misconceptions in Chemistry*, BestMasters, https://doi.org/10.1007/978-3-658-51032-9_5

that mistakes in solving equations are due to a lack of mathematical skills. However, students often demonstrate the necessary mathematical abilities, so the mistakes are more likely to result from a failure to grasp the true meaning of formulas and equations (Taskin and Bernholt, 2012), so they struggle to solve them correctly. A study by Heuer (2009) confirms this idea: Although students feel competent regarding chemical formulae in general, they find setting up and balancing reaction equations much more difficult and make many mistakes here.

Solving chemical equations, especially redox reactions, serves as the use case here, so the idea is to create an ITS which fulfills the requirements of making students understand chemical (redox) reactions correctly. A chemical equation is a written representation of chemical reactions consisting of symbols (such as "+" and "→") and chemical formulas with reactants on the left side of the reaction arrow and products on the right side. For example,

$$2H_2 + O_2 \rightarrow 2H_2O \tag{5.1}$$

is a chemical equation depicting the reaction that hydrogen and oxygen bond to water. The law of conservation of mass states that the number of atoms that are present in the reactants has to balance the number of atoms that are present in the reaction. Solving a reaction hence primarily consists of balancing the number and charge of the atoms. Now a redox reaction is any reaction involving the transfer of electrons between chemical elements. Next to balancing numbers and charges, it has to be ensured that the number of electrons absorbed is equal to the number of electrons discharged (Leifi-Physik, 2023).

Considering such chemical equations, three main problem areas explaining students' misconceptions can be distinguished: language, concept, and adequacy (Taskin and Bernholt, 2012). Language contains students' problems with the meaning and function as well as with the syntax of chemical formulae. For example, students tend to the oversimplified view that symbols and formulae are just abbreviations of the "real" substance names and are not linked to information on the composition of the substances. This suggests that students do not understand the submicroscopic level of the model.

Secondly, the concept refers to a conceptual misunderstanding. Students use subscripts and coefficients interchangeably or regard the reaction symbol as a mathematical equal sign (Taskin and Bernholt, 2012). They sometimes use formulae as an addition of all components, so they linearize formulae: There is an inadequate link between symbolic representations and actual concepts. For example, students simply go through the formula from left to right without mapping the particles. This also suggests an incorrect understanding of the plus sign ("+") between elements of the formula. Students interpret this plus sign mathematically, although in chemistry the reactants are not added together. Also, the atomic number and mass ratio are sometimes equated. Furthermore, indices in a formula are not related to the preceding element but to the following one. This demonstrates an incorrect understanding of the syntax rules of the formula language (Bernholt, 2021). Overall, the results indicate that students struggle with both the mode (cf. the coded signals) and the content (message as part of the conceptual semantics) of the formulae.

Thirdly, adequacy refers to the difficulties in interpreting and selecting formulae correctly, so the recognition of different types of formulas is often faulty. The differentiation between empirical formula and molecular formula is particularly prone to errors. However, as this master's thesis only focuses on chemical equations, especially redox reactions, only empirical formulas are of relevance here, so adequacy as the third area of concern will not be considered further.

To understand and address difficulties regarding redox reactions, Goes et al. (2020) conducted a (qualitative) systematic literature review of articles relating to teaching redox reactions that were published from 2000–2019. The articles reported several learning difficulties specific to redox reactions, such as understanding the electron transfer process, the assigning of oxidation numbers, chemical balance, and understanding redox reactions at the macro, micro, and symbolic levels. (Goes et al., 2020). Moreover, they reported most of the general learning difficulties regarding chemical formulae described above. However, the authors stressed that research on proposals aiming to overcome these difficulties is lacking.

How to overcome these difficulties? Generally, a systematic and confident use of chemical formula language in chemistry lessons should therefore be improved. As there are greater uncertainties when interpreting unfamiliar formulas and compounds, exercises that suggest similarities and differences between various examples are useful as a solution approach. In addition, formulas and particle concepts should be continuously linked. This counteracts memorization and instead induces an understanding of rules (Taskin and Bernholt, 2012), thereby counteracting both the difficulties arising from a faulty understanding of language and concept. Setting up reaction equations in general and redox equations in particular should be encouraged (Hansen, 2022). An ITS can be of great help here by requiring students to give step-by-step answers to unknown examples based on familiar examples. It also provides an easily accessible and intuitive opportunity to practice solving chemical equations.

5.2 Why Basic Pedagogical Concepts to Teach Chemistry are Fulfilled via Analogy

In chemistry, a distinction is made between (at least) two levels of observation: On the one hand the tangible macroscopic level, e.g. the actual substance water, which students can understand simply by observing. On the other hand, the sub-microscopic level is not accessible to the senses but requires a deeper understanding of the topic at hand. It refers to the molecule water, for example. A symbolic representation such as the formula language (H_2O) forms a bridge between the two levels: The sub-microscopic and simultaneously the macroscopic level is explicated via equations (Bernholt, 2021).

Consequently, to understand the formula language, a sub-microscopic model must also be learned, which considers the syntactic rules of the formula language to understand the formulas. This corresponds to a mental model of the meaning of the formula. An understanding of reaction schemes is therefore only possible if the general syntax of the structure is understood. The systematic formulation of formulas therefore involves the integration of the form of representation with relationships at a sub-microscopic level. As explained before, an analogy can provide such an indirect visualization of submicroscopic features, thereby enabling students to interpret these features correctly by mapping the analog to the target concepts.

A second aspect is that students should be familiarized with representations of the formulae and be able to deduce the spatial structure of a molecule. A formula should activate prior technical knowledge. This can be explained by the limited capacity of the working memory. To counteract this limitation, information must be combined into blocks or "chunks". Such integrated clusters of associatively linked information make it possible, for example, to have the entire rules of the formula language as a block in the working memory to solve a problem. Memorization, on the other hand, contradicts this chunking idea and is therefore not a sustainable strategy (Bernholt, 2021). Again, an analogy—especially based on WOE—ensures a true understanding of the underlying structure, so the students do not simply learn one example by hand but broaden their understanding and build chunks in their working memory.

Lastly, prior knowledge must be available for successful learning. In addition, the new knowledge must be assessed as relevant (and the relevance of the formula language is often not recognized). The new knowledge is then linked to prior knowledge, in the case of chemistry by explicitly linking symbols and representations to the underlying chemical concepts and principles (Taskin and Bernholt, 2012). As analogies help to consider the topic at hand as relevant and important and AWOE build on prior knowledge, analogies again seem a fruitful method to teach chemical equations.

Some studies support this claim. For example, both worksheet intervention models and tactile models are fruitful, well-established tools in educational practice, but reviews on students' problems with chemical formulae indicate that these tools are not sufficient to solve the problems in this domain (Taskin and Bernholt, 2012). On the other hand, Sarantopoulos and Tsaparlis (2004) conducted a longitudinal study in a naturalistic (classroom) setting to test if the use of chemical analogies enhances the cognitive and affective abilities of students and may therefore overcome the shortcomings of the methods described above. Their results provide evidence that long-term use of analogies aids teaching chemistry as the students performed better in their end-of-year final exams and considered the methodology used (the analogies) to be more interesting. The effect was greater for lower cognitive development students. Students fill knowledge gaps with several self-generated rules that are analogous to basic principles they already know. To support this claim, Cabubas and Mustacisa-Lacaba (2023) conducted a quasi-experimental study with grade 10 students to test the effect of pictorial analogies on their understanding of chemical equations. Comparing their pre- to post-test results on a multiple-choice questionnaire, the experimental group presented with pictorial analogies had a significantly higher learning gain than the control group.

Put together, several studies suggest that tutoring systems work well in chemistry and analogies are a fruitful tool to teach chemistry to students—so let us combine them.

5.3 Architecture

Regarding the architecture, the ITS consists of a user interface module, a feedback or tutoring module, a student model, a domain knowledge model, and a database of problems (see Appendix D in the electronic supplementary material). All problems and equations were designed and checked in cooperation with the "Institut für Pädagogik in den Naturwissenschaften" and hence with experts in the field of teaching chemistry. A main Python file ensures the application is running smoothly and the corresponding modules are called.

Additionally, my tutoring system roughly follows the Model-View-Controller design pattern which divides a given software application into three interconnected components (Freeman et al., 2008):

1. The **Model** holds data, states, and application logic, such as instructions or equations. It represents the core functionality and data of the application. The model notifies the view and the controller of any changes to its state.
   The **Model** in my implementation consists of the backend of Prolog rules.
2. The **View** depicts a representation of the model to the user. It visualizes the data that the model contains and generates output representation to the user based on changes in the model.
   The **View** in this thesis is represented by the web frontend.
3. The **Controller** takes user input, manipulates the model, and sometimes the view. It acts as an interface between the model and the view components, managing the data flow into model objects and updates the view whenever data changes.
   The **Controller** here is within the server.

But how do these modules interact to guide students in solving chemical equations? Before explaining the submodules in more detail in the next subchapter, the following tutoring system process flowchart (inspired by Myneni et al., 2013) depicts how the user interacts with the system (see Fig. 5.1). Such an activity diagram coordinates complex activities—in this case, the steps taken until an equation is solved. First, the controller starts the app, and a new equation is selected (randomly). The next step is then displayed to the student. The student answers and that answer is sent to the student module which forwards it to algorithmic debugging. The debugging calls the expert module which solves the given equation up to the given step based on its logic and rules. The debugger receives the expert answer, and checks if the two answers match. If so, the student module could not detect any errors and signals the interface module that the student's answer was correct, and the next step is displayed. In case of a wrong answer, the student module calls the feedback module which finds a suitable analogy, presents it to the expert module, receives its answer, and returns it to the frontend interface. Thereby, the analogy is shown to the student. The student then has to repeat the step. If all steps are solved correctly, a new equation can be chosen, or the tutoring ends.

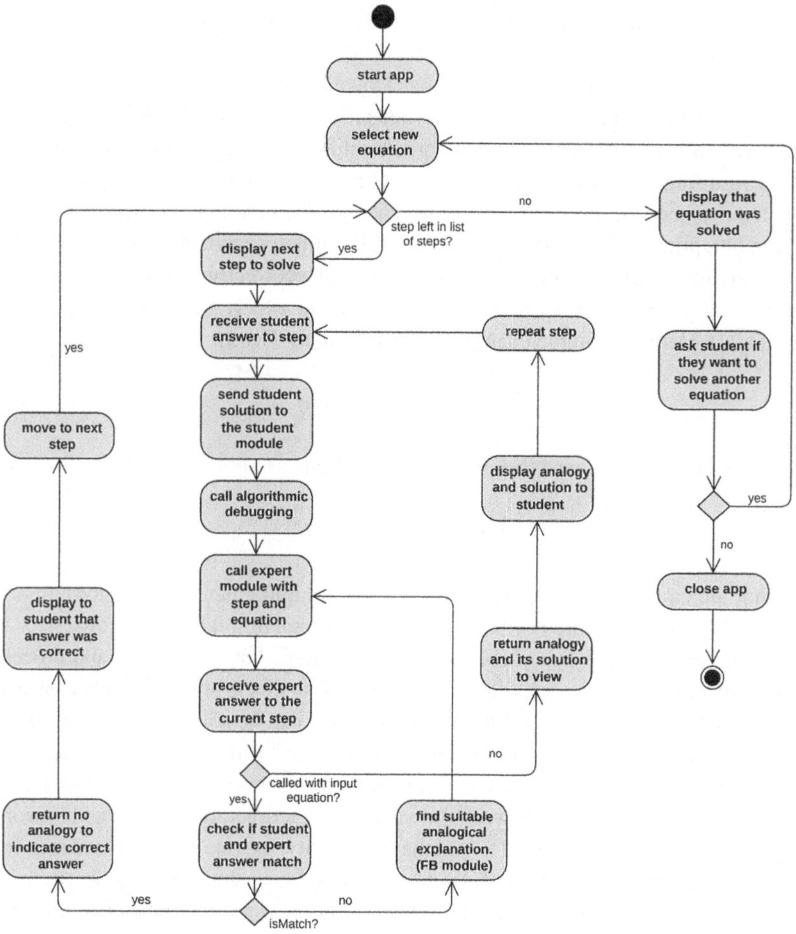

Fig. 5.1 Architecture of the ITS

5.4 The Process of Implementation

Before discussing the actual implementation of the architecture just described, a preliminary remark seems in order: What exactly is meant by solving a (redox) reaction? There are several typical sample items of redox reactions. Some of them are abstract problems formulated in text form (Textaufgaben). Such items are not part of the system, although they may be a useful add-on later (see Chap. 6). Other items require to give sensible solutions to an educt. For example, given Na_2 and O, what could the result of their reaction be? However, such questions firstly require a deeper understanding of redox reactions and are therefore not suitable for beginners, and secondly do not have only one correct solution. In many cases, elements can react in different ways, so there are several possible solutions and they do not fit the requirements of this tutoring system. The most common item is to balance out a given blueprint of a formula. It is very useful as it helps students to understand the steps involved and to develop an intuition on what a correctly solved redox reaction looks like (personal conversation with a chemical expert from the IPN (ChEx), 20.03.24). Here is a simple example of such an item: Given

$$\mathrm{Fe} + O_2 \rightarrow \mathrm{Fe}_2 O_3 \tag{5.2}$$

the student should balance the equation to result in

$$4\mathrm{Fe} + 3O_2 \rightarrow 2\mathrm{Fe}_2 O_3 \tag{5.3}$$

so that the equation makes sense. These items are implemented in the system and the underlying procedure is what is meant here by solving redox reactions.

There are several ways of solving redox reactions, but due to time and process constraints, I had to narrow it down to one technique that is both relatively easy and able to solve more complex redox reactions. They can be solved by inspection, by simply "looking" at the equation and going back and forth between products and reactants until everything is balanced out. However, as this strategy is not rule-based and thereby imprecise, it only works for very easy reactions and hence does not work as the tutoring strategy of an ITS. Another possibility, the one implemented here for redox reactions, is called the half-reaction method. Many students consider the half-reactions to be easier than solving the whole equation, especially when first learning about redox reactions (personal conversation with ChEx, 20.03.24). Furthermore, though the principal steps of solving redox reactions are the same, there are possible variations in the order and partial steps. I followed the steps from Leifi-Chemie (2024) as they are relatively straightforward and are applicable both to easy and more difficult reactions: To balance a redox equation using the half-reaction method, the equation is first divided into two half-reactions, one representing oxidation and the other one representing reduction. The equations for the half-reactions are then balanced for number and charge and, if necessary, adjusted so that the number of electrons transferred in each equation is the same. Lastly, the half-reaction equations are added together to provide a balanced equation of the redox reaction.

Importantly, however, while many easy redox reactions do not require explicitly searching for oxidation numbers, the expert module of the ITS always assigns oxidation numbers to the atoms or molecules. I decided to integrate them as many students struggle with this step, so it is important that the expert module "knows" how to assign the numbers in order for the feedback module to derive suiting analogies. Secondly, more complex equations require the use of oxidation numbers and as stated before, I want the system to be applicable to those reactions as well.

Balancing the charge is another essential part of solving redox reactions which needs further inspection. If a more demanding redox reaction is to be solved, this balancing may require adding H_3O^+ or OH^- molecules to the formula. Unfortunately, there are no general rules when to add H_3O^+ or OH^-, but it depends on the specific equation in question and its underlying chemical features. As this is both too complex and distracting to integrate into the tutoring system, a workaround is used. Each possible input formula is assigned an index number as an argument called 'Complexity', which can be either 0,1, or 2 (see Table 5.1). This index is given into the principal predicate `solving_redox/2` along with the formula and is used to balance the charge.

Table 5.1 Explanation of Complexity Indices

Index Number	Explanation
0	No additional balancing is necessary (easy input formula).
1	If needed, H_3O^+ should be added (cf. for an acid reaction).
2	Only OH^- may be added.

This allows for quite a bit of flexibility within the system: By requiring such an index, the system can not only solve easy and complex equations correctly, but it is also chemically precise and provides a single, unambiguous correct answer which can then be compared with the student answer.[1]

[1] If the system would allow to add both H_3O^+ or OH^- molecules, or, even worse, other charged molecule, it may at first seem easier to handle and, solely within the context of balancing charges, it is still correct. However, considering that an equation always represents a chemical formula which again represents subatomic chemical structures, the answer may be correctly balanced, but chemically absurd. Moreover, it would allow for several correct answers: Given that the educt has a charge of $+2$ and the product of -1, I could balance it out by either adding three H_3O^+ molecules, or three OH^- molecules, or two H_3O^+ and one OH^- molecule, etc.

Providing such a single correct answer is very promising as it is much easier to compare the answer given by the expert module to the answer given by the student if they match exactly. However, it requires certain conventions as to where chemical features are added. I will quickly mention these for the expert module, but the actual comparison happens in the tutoring module. As it is common in chemistry to add electrons after the redox partner, I decided to do the same. On the other hand, any additional molecules added to the formula are added in the beginning. Again, many students do the same. Otherwise, the ordering of the elements remains unchanged. For example, a formally correct but chemically incorrect reduction formula is

$$[(na, 2, 1, 0), plus, (e, 2, 1, -1), arrow, [(h, 3, 2, 0), (o, 3, 1, 0)], plus, (na, 2, 1, -1)]. \tag{5.4}$$

Any other ordering of the elements is considered formally incorrect.

Lastly, using the half-reaction method sometimes requires splitting a molecule into its atoms. For example, if oxygen and sodium react to sodium oxide (Na_2O), oxygen is part of the reduction and $Na2$ is part of the oxidation. This splitting works straightforward as there are no additional atoms in the molecule. However, some reactions require splitting molecules with more than two atoms. Take, for example, the reaction of sulfur dioxide and water to sulfurous acid (H_2SO_3). The sulfurous acid molecule consists of three atoms (or elements), namely hydrogen, sulfur, and oxygen. The question is which of these elements are part of the oxidation, and which belong to the reduction. Again, there is no general chemical rule to solve this issue as some elements, for example, sulfur, behave differently each time, depending on the other atoms and their indices given in the molecule. In other cases, such as in the example given above, the elements may belong to both oxidation and reduction. As such examples are difficult corner cases that are rarely taught in school at all, it makes sense to simply leave such cases out. The system thereby entails a restriction that given a molecule has to be split for the half-reactions and it is not a recombination-equation, no more than two atoms to such a molecule are allowed (see Chap. 6).

Put together, solving redox reactions consists of 6 rules (although for easy exercises steps 4 and 5 are merely test steps to ensure the previous steps worked correctly):

5.4.1 The Expert Module

To implement a system to solve redox reactions, a clear set of rules for each step on solving them was needed (see Table 5.2) and put these steps into the main predicate (see Appendix B in the electronic supplementary material). Following the rules, I implemented a step-based approach in the sense that each step of solving a formula should be tested. If a student assigns incorrect oxidation numbers, for example, the error is detected in this step and an analogous example of correctly assigned oxidation numbers is presented. After some preliminary remarks, I will lay out the implementation of each step in turn.

Table 5.2 Steps in Solving Redox Reactions

Index	Step
1	Assigning oxidation numbers
2	Detecting oxidation and reduction (the half-reactions)
3	Balancing the number of electrons
4	Balancing the charge
5	Balancing the number of all atoms of the educt and product.
6	Combining the half-reactions into the full solution.

Chemical Representation

Firstly, the chemical representation of formulas, atoms, and molecules had to be considered. An atom is represented as a four-digit tuple consisting of

$$(< \text{symbol} >, < \text{coefficient} >, < \text{index} >, < \text{charge} >)$$

so $2O_3^-$ is represented as $(o, 2, 3, -1)$. Molecules are lists of tuples of their atoms, so CO_2 is $[(c, 1, 1, 0), (o, 1, 2, 0)]$. A formula is then a list of sublists (molecules) and tuples (atoms), along with the chemical signs „plus“and „ow“which are known via an `is_sign/1`-predicate. The formula

$$\mathrm{Al_2O_3} + \mathrm{Mg} \rightarrow \mathrm{Al} + \mathrm{Mg^{2+}} \tag{5.5}$$

hence evaluates to

$$([[(al, 1, 2, 0), (o, 1, 3, 0)], plus, (mg, 1, 1, 0), arrow, (al, 1, 1, 0), plus, (mg, 1, 1, 2)]). \quad (5.6)$$

But how are atoms and molecules defined? Again, a molecule is defined in terms of its atoms by iterating over the list of atoms and checking if each atom is a valid chemical element. Any atom is hence an element with certain properties, denoted by the predicate `element/7` with

$$(< \text{atomic\_number} >, < \text{name} >, < \text{symbol} >, < \text{row\_number} >, < \text{column\_number} >, < \text{group\_type} >, < \text{electronegativity} >)$$

as arguments. For example "$element(1, hydrogen, h, 1, 1, 'a', 2.1)$" constitutes an element and thereby an atom. All necessary information to solve redox reactions is provided by the seven arguments of an element. For example, using $column\_number$ and $group\_type$, the group of each element can be deduced. Also, importantly, by this definition, an electron is not considered an element. This is not only true but also necessary when balancing the number or charge of all atoms (not the electrons) in an equation. However, an electron is represented as an atom, so as (e, <coefficient>, 1, -1).

Lastly, how to realize the charge had to be considered. It is easy for atoms as their charge simply is the fourth argument of the quadruple. In molecules, however, the atoms themselves have a charge which is then balanced with the other atoms, so that the overall charge of the molecule is usually zero. If a molecule has a charge, it is put after the last atom, cf. $2SO_4^-$. But it is obviously wrong to assign the charge of -1 to the last atom of the molecule (in this case, oxygen). Neither does it make sense to assign the individual charge the atoms would have if not balanced by the structure of the molecule as the atoms are part of a molecule. In agreement with ChEx, I decided to implement the overall charge of a molecule as the individual charge of each of its atoms. This is very close to the chemical truth and eases the following steps. Hence, in the given example, the molecule is represented as

$$[(s, 2, 1, -1), (o, 2, 4, -1)]. \quad (5.7)$$

Helper Predicates

Let me first mention some helper predicates needed for the sub-steps discussed later. Some introduce important chemical concepts, while others are mainly concerned with lists, so one can iterate over the equations or molecules. Next to the representation of atoms, molecules, and elements already discussed, `is_sign/1` defines both "+" and "→" as signs and is used to skip these items in an equation. Furthermore, the rules of oxidation numbers place importance on metals and halogens, so `is_metal/1` and `is_halogen/1`-predicates were needed. Other helpers use list iteration to retrieve the educt (first part of the equation) and product, append parts of equations together, search for an atom in a list, reduce or delete molecules, add an arrow between list elements, or change all coefficients of a molecule based on the input number. Another important predicate ensures that there are no sublists to a list, thereby "converting" molecules into atoms. Many of these predicates are needed because the built-in predicate `member/2` does not work given the atom-and-molecule structure. For example, one can check if an atom is a member of an equation or if it is a member of a molecule, but if it is within a molecule in the equation, `member/2` would return false. The predicate could hence only be used with electrons, which are not atoms and can thereby never be part of a molecule. The least common multiple (lcm) of two numbers is needed for balancing the atoms and electrons, so a predicate is provided. Lastly, the predicate `twice_difference/3` returns a triple of (<element>,<difference>,<type>), for example (na, 2, $reduction$) used to ensure the equation at hand is indeed a redox reaction.

Assigning Oxidation Numbers

The oxidation number is a measure of the degree of oxidation of an atom in a substance. The oxidation number is the real or hypothetical charge that an atom would have if all bonds to atoms of different elements were completely ionic. There are several fixed rules of different complexity on how to assign these numbers. I aimed to implement the easiest possible set of rules I could find, not only to ease the process of implementation but also to give more sensible rules as output if a student struggles to assign the oxidation numbers correctly. Furthermore, a consistent set of rules is required. That is, the proper application of the rules should always lead to the same correct result. To ensure this, priorities are often introduced (Leifi-Physik, 2023). I ensured such priorities via the natural reading of Prolog, thereby hoping to fulfill both the requirements of consistency and simplicity. Put together, the set of rules are listed in Table 5.3. Note that the sum rule is taken care of both via the rules for molecules and the `get_oxidation/2`-predicate, so it is commented out in the code. Metals (see rule 4) are deduced using the `is_metal/1` and the `element/7`-predicates.

Importantly, these rules imply said hierarchy: As Prolog rules are always traversed from top to bottom and from left to right, the first rule is applied first. For example, it is not always the case that the oxidation number of oxygen is −2—as

Table 5.3 Rules for Assigning Oxidation Numbers

Rule Number	Rule
1	The oxidation number of an uncharged atom is zero.
2	If a single atom has a charge, its oxidation number equals that charge.
3	The sum of the oxidation numbers of all atoms in a molecule equals their total charge (sum rule).
4	For all metals in the main group, their oxidation number equals their group number.
5	The oxidation number of fluorine in molecules is –1.
6	The oxidation number of hydrogen in molecules is –1.
7	The oxidation number of oxygen in molecules is –2.
8	The oxidation number of halogens in molecules is –1.
9	The more electronegative element within a molecule is given a negative oxidation number.

implied rule 7—but any of the exceptions in which the oxidation number differs are already taken care of by the rules stated above (for example, when oxygen reacts with a metal). As a second example, it should always be checked first if the charge of an atom is zero and, if so, assign zero as its oxidation number as this is the most basic rule. It is hence listed first. This implies that the ordering of the oxidation number rules is not to be changed or some errors might occur.

I included some simplifications. The first one was to not explicitly consider the formal charge of ions within a molecule for the assignment of oxidation numbers (but they became important later on). For example, in a $NaCl$ molecule, Na^+ and Cl^- ions are present, not the electrically neutral Na and Cl atoms. However, the molecule is balanced out as the ions balance each other out. But their charge must be counted as part of the oxidation state and, again, all charges must be balanced out in the end. The charge of ions could be neglected as it is implicitly part of the rules: The rules for molecules ensure that it is checked if any of its atoms have a charge (e.g. by belonging to a certain periodic group). It was tested for said $NaCl$ molecule represented as $[(na, 1, 1, 0), (cl, 1, 1, 0)]$ and for dichromate represented as $[(cr, 1, 2, 0), (o, 1, 6, 0)]$. In both cases, the system returned the correct oxidation numbers. Hence, as ionic atoms are implicitly taken care of in the rules used, I decided not to add them explicitly to reduce complexity.

The same holds true for electronegativity. Electronegativity is a chemical property that describes the power of an atom to attract electrons toward it and is one way of assigning oxidation numbers to elements. Although the implementation of the elements allows for a comparison of electronegativity using their $7^t h$ parameter (the element with higher electronegativity receiving a higher oxidation number), such a rule was left out at first.

However, when tested against the implemented test suite (see Chap. 6), the rules did not always return the correct solution as the approach was slightly oversimplified without the use of electronegativity. In certain cases, the main predicate was caught in an endless loop because it could not assign the oxidation numbers correctly: For any reaction of a metal with sulfur, electronegativity is crucial to assign the oxidation numbers correctly. For example, it could not find the oxidation numbers of iron sulfide. I hence added a new error condition if no oxidation numbers are found to avoid the endless loop and added a ninth predicate for oxidation numbers, namely `oxidation_number_electronegativity/2` which takes a list of elements as input and returns a tuple of the element with the highest electronegativity as well as the corresponding oxidation number of that element. This way, the more electronegative element within a molecule is given the negative oxidation number as this element aims to accept new electrons. If the element is a member of the main group, its oxidation number simply is the number of electrons it misses up to the noble gas configuration in group 18. Take sulfur, for example, which belongs to group 16 and is hence in the main group: Its oxidation number is $16 - 18$, so -2. To figure this out, the predicate firstly calls `find_electronegativity/2`, which retrieves the electronegativity of each element in the equation from the `element/7`-predicate (remember that the last argument of this predicate simply is the electronegativity). The predicate then calls a second helper, namely `max_tupel/2`, which iterates over a list of tupels consisting of (<element>,<number>) to find the highest number—in our case, the highest electronegativity. To ensure this element is called, a new `get_oxidation_member/4`-predicate is needed in case the previous predicates—and therewith, the other rules for assigning oxidation numbers—fail. Recall that this helper only considers molecules as within molecules the electronegativity is relevant. This predicate calls the `get_oxidation_helper/2`-predicate on a molecule (which is a list), updates the counter, adds the returned element and its oxidation number to the oxidation list, and recursively calls `get_oxidation_helper/4` on the remaining elements of the molecule. Importantly, it also calls `remove_from_mol/3` to ensure the element for which the oxidation number is now assigned is removed from the molecule. By adding these predicates, the system now works as intended and the testing was successful.

But how are these rules used within the system? The main predicate for the oxidation numbers is called `find_oxidation_number/7` which receives a formula as input and returns a list with all elements within this formula and its respective oxidation number (B). For example, given the formula

$$[[(al, 1, 2, 0), (o, 1, 3, 0)], plus, (mg, 1, 1, 0), arrow, (al, 1, 1, 0), plus, (mg, 1, 1, 2)] \tag{5.8}$$

it returns

$$[(al, 3), (o, -2), (mg, 0), (al, 0), (mg, 2)]. \tag{5.9}$$

This is done via calling the `get_oxidation/2`-predicate which iterates over an input list (an equation), skips any signs (so "+" or "→"), and calls the `oxidation_number/2`-predicate for atoms or the `oxidation_number-molecule/2`-predicate for molecules. More precisely, in the case of a molecule, a helper predicate is used to iterate over the list of atoms within the molecule to ensure an oxidation number is assigned to each of the atoms within the molecule. The oxidation_number-predicates return a tupel of (<element>,<oxidation_number>), which is appended onto the oxidation list. A correct ordering of this list is ensured via calling `order_molecule/3` which is necessary for tracing. Next, another helper predicate called `no_sublists/2` is called which takes the oxidation list as input and deletes any sub-lists resulting from appending the atoms of molecules (so elements of a sub-list) separately. This final oxidation list is then returned. Now that we have an oxidation list, it is necessary to separate the reduction and the oxidation. Hence, the next step iterates over the oxidation list to check if two elements differ in their oxidation number and, if so, if one of these emits and the other absorbs electrons. The `findall/3`-predicate is used here to check for any element with different oxidation numbers that appears twice in the equation. The corresponding predicate `twice_difference/4` returns a triple of (<element>,<difference_in_oxidation_number>,<oxidation_type>). Following the definition of a redox reaction given above, if two such elements exist and one of these elements' oxidation number decreases while the other increases, the equation in question is indeed a redox reaction. For example, it may return $(o, -2, reduction)$. Using the example above, the findall-request

$$findall((X, Y, Z), twice\_difference(X, [(al, 3), (o, -2), (mg, 0), (al, 0), (mg, 2)], (X, Y, Z)), RedoxList). \tag{5.10}$$

returns

$$RedoxList = [(al, 3, reduction), (mg, 2, oxidation)]. \tag{5.11}$$

To ensure the reaction is a redox reaction, the predicate now checks that both a reduction and an oxidation are part of the redox list and that the oxidizing and the reducing elements differ.[2] Both the elements and their factors (that is, the difference between the oxidation number of the redox partners in the educt compared to the product) are stored in variables. Another error predicate returns an error if these requirements are not fulfilled, so if the input formula is not a redox reaction. Otherwise, both the oxidation list and redox list as well as the elements of reduction and oxidation, along with their factors, are returned.

Setting up the Half-Reactions

As it was ensured that the equation in question is a redox reaction, the half-reactions are now set up using `make_reduction/8` and `make_oxidation/8` which have the same structure. They take the equation, the oxidation and redox list, and the reduction or oxidation element along with its factor as input to return a split value and the half-reaction. Firstly, the predicate `make_halfreaction/7` is called both for the reduction and oxidation, which takes as input the redox partner, the formula, and the oxidation list, and returns the corresponding (incomplete) half-reaction of reaction partners, the count of the redox partner in the equation as well as the index number "*Split*". Importantly, this predicate ensures the correct representation of all redox partners. This especially includes splitting a molecule into its atoms, if both atoms are a redox partner. After splitting the molecule up into atoms using `no_sublists/2`, the atom in question (the oxidation or reduction partner) is represented correctly by replacing its old coefficient with the product of this old coefficient and its index. The oxidation number of the element becomes the charge. This way, Na_2 within a molecule is now—as an atom—represented as $2Na^+$. Additionally, the Split attribute is assigned here: If no splitting takes place, the number is zero. If the educt is split, the number is 1 and if the product is split, the number 2 is returned. If an invalid representation is found so the molecule has to be but cannot be split, a split value of 3 is returned to indicate an error. This Split attribute becomes important later on when combining the half-reactions. Using this attribute, it is easier to figure out if atoms have to be made a molecule again. For example, this predicate may return $[(na, 1, 1, 0), (na, 2, 1, 1)]$.

Next, an arrow is added to distinguish the educt from the product. An add-one `check_count/5` is called to change the split value to four if an element occurs more than twice in the equation. Then the number of electrons that are transmitted is added. To do so, the difference in oxidation number of the redox partner has to

---

[2] Note that this is not true in the case of disproportionist reactions, but as these are left out, it is the easiest way to ensure both half-reactions can be set up correctly.

be multiplied by its index and coefficient on the side of the equation where the electrons are added. For example, in the case of our sodium example, the difference in oxidation number is 1, and, as it is the oxidation, the electrons will be added to the product of the half-reaction. This means that each sodium atom will accept one electron. We hence consider the index and coefficient of the sodium atom in the product to know how many atoms there are. The coefficient is 2 and the index is 1. There are hence $2 \times 1 = 2$ atoms each accepting one electron, so the overall calculation of $1 \times 2 \times 1$ yields that 2 electrons have to be added to the product. As electrons always have an index of one, this factor is simply their coefficient. After this final step, the full half-reaction is established and returned. Given the example of sodium and oxygen reacting,

$$[(na, 1, 1, 0), plus, (o, 1, 2, 0), arrow, [(na, 1, 2, 0), (o, 1, 1, 0)]] \tag{5.12}$$

both the arrow and 2 electrons would be added to the oxidation, so the equation

$$[(na, 1, 1, 0), arrow, (na, 2, 1, 1), plus, (e, 2, 1, -1)] \tag{5.13}$$

is returned.

Balancing Redox Partners and Electrons

Now the half-reactions are set up, they have to be balanced. This basically means that the amount of something on one side of the equation should equal the amount of the same thing on the other side of the equation. There is no real "rule" here as to in which order this takes place. It is perfectly possible to firstly balance electrons, then atoms, or even to start with the charge. However, based on personal conversations, firstly balancing atoms that are either reduction or oxidation partner seems the most common procedure. I hence implemented this step first. The corresponding predicate, `balance_atoms/4` takes a half-reaction formula and its split value as well as the redox partner as input and returns the half-reaction formula with a balanced number of atoms. It is called both for the reduction- and the oxidation-equation to balance them out. If the split value does not equal 4, the `find_atom/3`-predicate is called on both the educt and the product which returns the representation of the atom as given within the formula. For example, using our running example, calling the `find_atom/3`-predicate with the sodium-product and *na* as X, it returns $(na, 2, 1, 1)$. Both the index and the coefficient of this returned atom are stored in variables, because the number of the atom in the product is, of course, the index multiplied with the coefficient. This way, the number of the redox partner in both educt and product is found and given into the helper predicate calculate_balance/7 which takes these numbers, the educt and product, the element as well as the whole formula as input and returns the balanced formula. If the number of the atom in educt and product is the same, it is already balanced, so the formula is returned unchanged. If, however, they differ, the least common multiplier (lcm) has to be found and divided by the respective number to figure out the multiplier with which to update the elements. As an example, if three oxygen-atoms are given in the educt and two in the product, the lcm is 6. Now 6 divided by 3 is 2, so the number of oxygen-atoms in the educt has to be multiplied by 2, whereas the number in the product is multiplied by 3 ($\frac{6}{2}$). Importantly, when updating the atoms, only the redox partners are updated. This means we do not update all elements of the equation, but simply our oxidation or reduction element along with the number of electrons: Recall that each element emits or accepts a defined number of electrons. Hence, if the number of atoms increases, the number of electrons has to increase by the same factor. To ensure this, the predicate update_element/4 is called which updates the coefficient of the input element in a given equation by a given multiplier. The predicate is called both on the redox partner and on the electrons, but not on other elements in the equation. It is ensured that if the searched for element is not found in the formula, the predicate simply returns the formula unchanged. This is important as, for example, the product of a reduction does not contain any electrons. Otherwise, the balanced formula is returned. In our running example, the number of sodium would thereby be balanced to two atoms per side, so the returned formula is

$$[(na, 2, 1, 0), arrow, (na, 2, 1, 1), plus, (e, 2, 1, -1)]. \tag{5.14}$$

Next in line, the number of electrons between oxidation and reduction have to be balanced as well. Again, the definition of a redox reaction requires that the same number of electrons is emitted and accepted, so balancing is needed here. The predicate balance_electrons/4 takes both oxidation and reduction equations as input and returns both with balanced electrons. Using the member/2-predicate, the number of electrons in the oxidation and reduction formulas is retrieved and, again the lcm of the two numbers is calculated. The lcm divided by the number of electrons gives us the multiplier with which to update the coefficients. But this time, as the number of electrons defines the whole half-reaction, all atoms and molecules of the formula have to be updated, not just the redox partners. Consequently, the predicate update_formula/3 is called which updates the coefficients of all items of a given formula with a given multiplier using list iteration. In our sodium-example, the reduction formula has 4 electrons, but the oxidation only 2. The entire oxidation formula hence has to be multiplied by 2 ($\frac{4}{2}$) and is returned as

$$[(na, 4, 1, 0), arrow, (na, 4, 1, 1), plus, (e, 4, 1, -1)]. \quad (5.15)$$

Balancing the Charge

As the fourth step, the charge has to be balanced. However, for many formulas, this is not necessary. If no element of the equation other than the redox partners have a charge, the charge is already balanced. Put differently, many easier equations do not have any elements other than the redox partners. Their charge is balanced by the previous step: If the same number of electrons is emitted and accepted, the charge has to be balanced. But some formulas have additional charged atoms which are not yet balanced. Depending on submicroscopic features of the equation, H_3O^+ or OH^- are added. To avoid the difficulties arising from the decision on which molecule to add, the complexity index C is added here. The index is part of the main predicate and now called in this step. If $C = 0$, the simple case is given where there are no elements other than the redox partners. In this case. The predicate `balance_charge/3` simply returns the input formula. In our running example, C is 0 because the oxidation equation only consists of sodium and electrons, which are the redox partners. The formula would hence be returned unchanged as

$$[(na, 4, 1, 0), arrow, (na, 4, 1, 1), plus, (e, 4, 1, -1)]. \quad (5.16)$$

If the formula is more complex, `balance_charge/3` calculates the charge of the educt and product by iterating over each atom and molecule and adding their charge to the overall counter. The difference between the charge of the educt and the product is given as input into the helper predicate `balancing_charge/6`. If the difference is negative, it means that the educt has more negatively charged elements. If $C = 1$ and H_3O^+-molecules are to be added, the predicate hence ensures that they are added to the educt. The number of molecules to be added equals the difference in charge between educt and product. If $C = 2$ and OH^- is added, it is appended to the product list. On the other hand, if the difference is positive, either OH^- is added to the educt (if $C = 2$) or H_3O^+-molecules are added to the product (if $C = 1$).

Balancing all Atoms

Next, the number of all additional atoms—those that are not redox partners—are balanced. Again, this is only necessary if we are dealing with a complex formula. Once more, therefore, our running example with $C = 0$ skips this step as the predicate `balance_additional_atoms/3` simply returns the input formula (see 5.16) if C is zero.

If, however, a molecule was added in the previous step, there are a number of H and O-atoms that are not yet balanced. This is done via adding H_2O-molecules in chemistry. Note that this requires a 2 : 1-proportion of $H - to - O$-atoms. For example, a difference of 4 H-atoms to 3 O-atoms cannot be balanced using H_2O. I used this as another indicator that all previous steps were done correctly. If, however the difference of H-atoms is not twice as large as the difference in O-atoms, the system throws an error and returns an empty list. Such an error may not only result from an error in previous steps, however. When coding the expert module, I noticed that some of the examples given in chemistry textbooks are actually faulty. The chemistry expert confirmed this observation. I considered to sort out such examples in the beginning. However, in the future, if the system allows the student to enter an equation, this is not enough. The system has to be able to deal with inconsistent, complex equations. As I noticed that in such examples, there is no 2 : 1-proportion of hydrogen to oxygen, I believe throwing the error described here greatly helps to cover the input of erroneous equations.

Putting these ideas together, the predicate `balance_additional_atoms/3` takes a formula and C and—if C is not zero—returns a balanced version of the formula. It calculates the number of H- and O-atoms in both educt and product via updating a counter when iterating over the input list, stores the difference of H and the difference of H atoms in variables, ensures the difference of H-atoms is twice as large, and calls the helper predicate `balancing_additonal_atoms/5`. This helper balances the equation by adding H_2O-molecules to the side with less oxygen and hydrogen atoms if C is not zero. For example, given a negative difference implies that there are more atoms in the product, so H_2O has to added to the product. The coefficient (or the number) of molecules is given by the difference in oxygen: If 3 O-atoms are missing in the educt, 3 H_2O-molecules have to be added.

Combining the Half-Reactions

Lastly, the two half-reactions have to be put together into one result formula that is then returned by the main predicate of the expert module. Several steps are needed here: Firstly, the educts and products have to be appended to form one equation; we secondly have to delete the electrons, as they are no longer needed as a helper structure to grasp the underlying chemical idea; thirdly, all elements that are not redox partners and occur on both sides of the equation may be reduced; lastly, if splitting took place, the atoms need to be combined back into a molecule with an updated coefficient. To do so, the predicate `make_fullreaction/6` takes the original formula as input, the oxidation and reduction equations, and both indices (Complexity and Split) to return the final formula. It firstly appends all educts and products of both reduction and oxidation calling the append-helper-predicates into one educt and one product. The ordering of the educt and products is then checked against the input equation to ensure the same structure (see below). Next, the number of electrons in this educt and this product is calculated and compared. Again, I checked that the number of electrons is the same to ensure no error has been made previously. The `delete_element/3`-predicate is called twice with the electron, once for the educt and once for the product, to return two formulas without electrons. Our sodium oxidation is hence combined with the corresponding reduction and deprived of its electrons. The new product is $[(na, 4, 1, 1), plus, (o, 2, 1, -2)]$.

Using these, the `reduce_reaction/5`-predicate reduces double-figure molecules that are in both the educt and the product if C is not zero. If C=1, occurences of H_3O^+ or OH^- if $C = 2$, are checked as well as H_2O-molecules in either case. Given the number of each molecule in the educt and product, another helper predicate is called. This helper predicate ensures that if there is a higher occurrence of the molecule in one part of the equation, this occurrence is reduced by the difference while the other molecule is deleted. If the numbers are the same, both molecules are deleted. For example, if there are 3 H_2O-molecules in the educt and 5 H_2O-molecules in the product, the predicate ensures that all molecules are deleted in the educt and $5-3 = 2$ H_2O-molecules remain in the product. The same is done for H_3O^+ or OH^-, so the predicate is called twice. These reduced educts and products are returned.

Now one final step is needed before calling `append_helper/3` again to combine educt and product into the return predicate, namely checking for the Split index. The predicate `add_input_molecule/6` takes the original formula as input to check how the molecule was formed there, the educt, product, and Split value, to return an updated educt and product. If $S = 0$, nothing is changed for no splitting took place. If $S = 1$, the educt is changed, and if $S = 2$, the product is updated by replacing the atoms in question with the molecule. Simply put, the atoms are made a list—and thereby a molecule. Their charge is now zero as the charge within a molecule is balanced out by its atoms, so the overall charge of a molecule is always zero. The coefficient is calculated by dividing the coefficient of the atom by the index number of that atom in the original molecule. Recall that the atoms were created by putting their index as the new coefficient, so to reverse this effect, their current coefficient has to be divided by the index the atom possesses in a molecule. Furthermore, it is ensured that the coefficient is the same for all atoms as all members of a molecule share the same coefficient. To further illustrate, the product of the sodium and oxygen reaction now reads $[[(na, 2, 2, 0), (o, 2, 1, 0)]]$.

The Main Predicate

The main predicate of the expert module, which is passed onto the algorithmic debugging, is called `solving_redox/2`. It combines all steps previously discussed (see Fig. 5.2. Importantly, all sub-steps predicates can be called separately as well, which comes in handy for the tutoring strategy: If an incorrect student answer is given, each step of the main predicate can be called one after another to check from with step the error originated and to present a fitting analogue. The `solving_redox/2` predicate firstly assigns oxidation numbers, makes two half-reactions, balances their atoms based on the split value (note that, for simplicity, that was left out of 5.2), balances electrons, charge, and additional atoms (if applicable), before combining the half-reaction into the solution, again taking the split value into account. The result of this last step, the `make_fullreaction/6`-predicate, is returned as the "SolutionFormula":

$$[(na, 4, 1, 0), plus, (o, 1, 2, 0), arrow, [(na, 2, 2, 0), (o, 2, 1, 0)]]. \quad (5.17)$$

Importantly, the expert module misses a final check to control if every atom is indeed balanced in the return formula. Neither `make_fullreaction/6` nor `solving_redox/2` provides such a thorough checking. Such a step is, however, recommend for students when solving redox reactions. For example, Leifi-Physik (2023) considers this the final step to take. It is important to not only check if their solution is correct, but also (potentially) to understand why it is. But to me it did not seem fitting for the expert module: After all, this step is only necessary if an error occurred in one of the previous steps. Considering the system works as intended, this cannot happen. And several checks are included to rule out small errors and to prevent next steps from happening if such an error is detected. Nevertheless, a thorough check-up could be added in the future, but is not added here, as it is fruitless to double-check for arithmetic errors a computer simply does not make.

5.4.2 Error Messages

Again, providing useful error messages is mainly part of the tutoring module. However, some error handling is included in the expert module to ensure the system works as intended. Some of these were mentioned in the previous subchapters. The errors are mainly concerned with ensuring the balancing of elements is even possible. There are some "wrong" formulas where the numbers are so incongruent that the steps provided to balance the equation do not work. Such formulas are either chemically wrong or extremely complex and hence never taught in school. For example, to counterbalance additional atoms with water, there have to be twice as many hydrogens as oxygen atoms present. Also, the number of atoms within a molecule to be split has to be two. If this is not the case, the formula may in theory still be solvable, but not for students. Considering the level of complexity

```
Input : equation, complexity
Output : balanced_equation
Function solving_redox(equation, complexity):
    // 1. assigning oxidation numbers
    for elem in equation do
        no ← find_oxidation_number(elem)
        oxidation_list.append(no)
    for elem in equation do
        if elem is in equation twice then
            if oxidation_list[elem1] > oxidation_list[elem2] then
                redElem, redNo ← elem, number of elem (index*coefficient)
                redDiff ← oxNu(elem1) - oxNu(elem2)
            else if oxidation_list[elem1] > oxidation_list[elem2] then
                oxiElem, oxiNo ← elem, number of elem (index*coefficient)
                oxiDiff ← oxNu(elem1) - oxNu(elem2)
    // 2. making half-reactions - with () marking the amount of that element
    oxidation ← ((oxiNo1)oxiElem1 → (oxiNo2)oxiElem2 + (oxiDiff×oxiNo2)electron)
    reduction ← ((redNo1)redElem1 + (redDiff×redNo1)electron → (redNo2)redElem2)
    // 3. balancing the amount of redox partners
    O ← kgV(oxiNo1, oxiNo2)
    R ← kgV(redNo1, redNo2)
    newOxi ← ((O/oxiNo1)oxiElem1 → (O/oxiNo2)oxiElem2 + electrons
    newRed ← ((R/redNo1)redElem1 + electrons → (R/redNo2)redElem2)
    // 4. balancing the electrons
    factor ← kgV(noElectronsRed, noElectronsOxi)
    multiplierR ← factor/noElectronsRed
    multiplierO ← factor/noElectronsOxi
    betterOxi ← multiply each element of newOxi with multiplierO
    betterRed ← multiply each element of newRed with multiplierR
    // 5. balancing the charge
    if complexity ≠ 0 then
        for each reaction in [oxidation, reduction] do
            charge1, charge2 ← get charge from educt (charge1) and product (charge2)
            if charge1 > charge2 then
                append(diff(charge1, charge2)×H3O+) to product or append(diff×OH−) to educt
            else if charge1 < charge2 then
                append(diff(charge1, charge2)×H3O+) to educt or append(diff×OH−)to product
    // 6. balancing all atoms
    if complexity ≠ 0 then
        for each reaction in [oxidation, reduction] do
            diffO ← (noOEduct - noOProduct)
            diffH ← (noHEduct - noHProduct)
            if noOEduct > noOProduct and noHEduct > noHProduct then
                product.append(diffO×H2O)
            else if noOEduct < noOProduct and noHEduct < noHProduct then
                educt.append(diffO×H2O)
    // 7. combining the half-reactions
    combined ← put betterOxi and betterRed into one equation, adapt ordering from equation
    balanced_equation ← delete all electrons from combined
    if complexity ≠ 0 then
        balanced_equation ← reduce H3O+, OH− and H2O from balanced_equation
    return balanced_equation
```

Fig. 5.2 Algorithm for Solving Redox Reactions

taught in school, such equations are not solvable. The system hence throws an error message in such cases and returns an empty list (or an empty equation), indicating that no solution can be found to such examples. The error messages are depicted using the built-in `write/1`-predicate in Prolog which writes the input term to the current output stream. The tutoring module takes this to mean that the input is faulty, reports it to the view and, correspondingly, requests another input sample. Lastly, some of the examples provided in the test suite are not redox reactions but other types of chemical reactions. Consequently, the `find_oxidation_number/6`-predicate returns false as it cannot find redox partners for oxidation or reduction and, therewith, the main predicate returns false. Any additional handling of these special cases is then given to the frontend as this is the only case where the main predicate simply returns false without any error given. This return hence implies that the input equation is not a redox reaction at all.

Additionally, the code of the expert module was tested, of course. Each predicate was tested in itself with different input types of both normal and corner cases. I also used test suites for black-box testing of the entire code along with some low-level white-box testing based of tracing examples (see Chap. 6).

5.4.3 Adjustments

Firstly, some minor mistakes occurred. For example, some more elements had to be added to the list of relevant chemical elements, such as tellurium. Some error cases had to be refined, especially as in some cases the correct error message was shown, but the system still continued processing with wrong answers or empty lists, often resulting in endless loops (see below). The number of electrons was calculated incorrectly, as it did not take the coefficient into account and the case where the corresponding element was not part of a molecule was forgotten. The latter resulted in a zero-division-error when calling `balance_electrons/4`. Both errors were fixed by adding a new case and multiplying the charge times the coefficient of the electrons. The charge of a molecule was represented incorrectly in `get_oxidation_helper/4`. Whereas the charge of each atom has to be added to the full charge (to ensure balancing), the overall charge of a molecule is applied to each atom within the molecule. This means that in case of a molecule, the overall charge simply equals the charge of the first atom within the molecule, but not the sum of the charge of all atoms. This was taken into account.

If each step of the system is executed correctly, it is not necessary to double-check the balancing of atoms. Indeed, the system always returns the correct answer without such an extra step. However, it is recommended as a test step to check that the number of each atom in the educt equals the number of the same atom in the product as this helps students to further understand why the previous steps worked out well. For example, Leifi-Chemie (2024) provides a table listing all occurrences of all elements as an overview. To me, this seemed much more reasonable than adding another predicate to the expert module which does not bring any progress. Hence, no further balancing of atoms is implemented here.

Furthermore, the ordering was faulty. There is no rule whether the first, second, third, etc. element of an equation is part of the oxidation or reduction, so it is difficult to know how to add the half-equations together correctly. This is an issue humans do not face as they would simply add the coefficients and additional elements, thereby ensuring the ordering is the same as in the original formula. However, as the implementation rebuilds the solution formula from scratch, a different ordering may occur. For example, upon the input listed below, the predicate correctly returns the equation for R:

$$\begin{aligned} &solving\_redox(([[(cr, 1, 2, 0), (o, 1, 3, 0)], plus, (al, 1, 1, 0), arrow, \\ &\quad (cr, 1, 1, 0), plus, [(al, 1, 2, 0), (o, 1, 3, 0)]], 0), R). \end{aligned} \tag{5.18}$$

$$\begin{aligned} R = {} &[(al, 2, 1, 0), plus, [(cr, 1, 2, 0), (o, 1, 3, 0)], arrow, \\ &[(al, 1, 2, 0), (o, 1, 3, 0)], plus, (cr, 2, 1, 0)]. \end{aligned} \tag{5.19}$$

However, the ordering is not the same as in the input formula, so when comparing this against the solution in the correct order, it does not return true.

I originally intended an `ordering/4`-predicate which iterates down the input equation until it is empty and starts adding content to the output equation. This way, however, the entire equation is built from scratch where every element is checked against the input equation, which is not very efficient. Instead, the predicate `put_in_order/4` takes two formulas as input (for example, the two educts of oxidation and reduction) as well as the corresponding part of the formula of the input equation (that is, the input educt) and returns the full balanced educt. The idea was to simply check if the first atom or the first element of the first molecule of the input formula is a member of the first formula. If so, that formula (cf. representing the oxidation) has to appear first in the solution, so `append_helper_plus/3` ensures to append the first formula, then the second. If, however, the `member/2`-predicate fails, it has to be the other way around, so the second formula is appended first onto the solution. This is a much simpler way as the formulas are simply appended, not rebuilt, while the correct order is ensured: The first element in the input equation has to be the first element of the output equation. Consequently, when tested against different cases, it always returned the correct ordering. In the example above, the main predicate now returns

$$R = [[(cr, 1, 2, 0), (o, 1, 3, 0)], plus, (al, 2, 1, 0), arrow,$$
$$(cr, 2, 1, 0), plus, [(al, 1, 2, 0), (o, 1, 3, 0)] \quad (5.20)$$

which is correct.

To ensure a correct order also for the more complex equations, it has to be ensured that OH^-,H_3O^+, and H_2O always appear at the beginning of an equation. The `further_ordering/3`-predicate therefore takes a complex formula as input and returns it with these molecules ordered to the front. It calls `put_further_in_order/4`, which takes a formula, the complexity (which may not be zero), and the molecule it should put in front as input and returns an equation sorted for that specific molecule. The predicate is hence called twice, firstly for OH^- or H_3O^+ (depending on the complexity value), then for H_2O, so H_2O is always at the head of the list. The predicate `put_further_in_order/4` checks how often the molecule is present in the given formula (cf., the product). If it is not a member, it simply returns the given formula. If it is a member once, it is moved up front and deleted from its previous position within the formula. If it is a member twice, meaning it was added both in the oxidation and in the reduction, these occurrences are added together and put as head to the output before both other occurrences of the molecule are deleted. This concludes to combine the half-reactions.

Further adjustments were needed to ensure a correct solving of more complex solutions. In case of these complex solution, an additional balancing of elements that are not redox partners is needed. One of the corresponding predicates, namely `calculate_charge/3`, was faulty. It calculates the charge of a given formula via list iteration and uses full charge as a counter which is then returned. It usually returned a charge of -1 because of the electrons as I forgot to consider the index and the coefficient, always resulting in one OH^- or one H_3O^+ molecule to counterbalance. However, the full charge of an atom or an electron is really given by multiplying its charge with the index and the coefficient. As mentioned above, the charge of a molecule is represented differently: Only the first atom within the molecule is considered and the molecule charge is simply the charge of the atom times the coefficient—as each atom of a molecule represents the full charge of one occurrence of that molecule. These adjustments ensured a correct ordering of all complex predicates. Adding these improvements, the predicate now works as intended.

Other errors were caused by the predicate `reduce_reaction/5` which aims to reduce any two-figured molecules that were added for complex equations. This predicate often failed as cases where, for example, OH^- is part of the educt but not of the product were missing. However, this is perfectly possible. Consider an equation where OH^- only has to be added in the educt of the oxidation. The molecule will then not appear in the product, even if in the reduction OH^- was also added to the educt. I hence broadened the predicate by including cases where OH^- is a member only of the educt or the product. In this case, no reduction is needed, so the input formula is simply returned.[3]

Finally, in the case of magnesium and hydronium ion reacting to positively charged magnesium and hydrogen, the resulting solution formula is correct, but not yet reduced: The correct, reduced solution is depicted here.

$$[(mg, 1, 1, 0), plus, [(h, 2, 3, 1), (o, 2, 1, 1)], arrow, \\ [(h, 2, 2, 0), (o, 2, 1, 0)], (mg, 1, 1, 2), plus, (h, 1, 2, 0)] \tag{5.21}$$

However the system returns this equation:

$$[(mg, 3, 1, 0), plus, [(h, 6, 3, 1), (o, 6, 1, 1)], arrow, \\ [(h,6, 2, 0), (o, 6, 1, 0)], (mg, 3, 1, 2), plus, (h, 3, 2, 0)]). \tag{5.22}$$

The whole formula is reducible by the factor 3. This is not chemically wrong; however, it equals an unreduced mathematical equation such as $3a + 3b = 6c$, which is not an incorrect, but a somewhat incomplete solution. As the case only occurred in this one formula, it was considered superfluous to add such a reducing-predicate for the moment. It is, however, a sensible expansion to the system in the future (see Chap. 6).

Lastly, although all answers returned by the expert module are correct, the half-equations may be faulty. Upon further checking that all sub-steps are correct as well, I discovered a problem with all recombination equations as the half-reactions are not fully balanced here. More precisely, those atoms "switching" partners in the recombination simply cannot be balanced in a half-reaction. This is hence not a shortcoming of the system, but of using half-reactions as a pedagogical helper method. For example, consider the reaction of copper oxide and iron to copper and iron oxide:

$$CuO + Fe \rightarrow Cu + FeO \tag{5.23}$$

[3] It may be necessary to add two occurrences of OH^- in the educt together. As described above, this is ensured by the ordering predicates.

The oxygen cannot be balanced in the half-reactions here. Consider the following half-reactions the system returns:

$$\text{RED: } CuO + 2e^- \rightarrow Cu \tag{5.24}$$

$$\text{OXI: } Fe \rightarrow FeO + 2e^- \tag{5.25}$$

Now the oxygen is not balanced, and it can only be balanced by adding a "new" element, namely oxygen, to the product in the reduction. This does not make sense chemically. Furthermore, the half-reactions work well enough, and the correct result is returned, so it seems complicated to change them. Upon discussing this issue, we decided to leave the expert module as it is (as the half-reactions are somewhat correct), but rather highlight such elements in the frontend to make sure students understand that these elements are neither part of the oxidation nor the reduction, but simply perform another chemical reaction at the same time (see Chap. 6; personal discussion with ChEx, 21.04.24).

5.4.4 Further Adjustments Regarding Difficult Reactions

Some difficult types of reactions have to be considered more closely: One problem concerns dis- or comproportionist reactions where an element both reduces and accepts electrons, so both oxidation and reduction happen within the same element. Such reactions are an exception to common chemical rules as many of the ordinary steps—and therewith, many of the steps I implemented—do not fully apply. For example, in the formula

$$CO_2 + C \rightarrow 2CO \tag{5.26}$$

carbon is redox partner both of oxidation and reduction. This contradicts a condition within `find_oxidation_number/6`, stating that the oxidation element may not be the same as the reduction element. The product molecule is also part of the reduction and the oxidation formula. Again, this contradicts common cases and my implementation as a single two-figured molecule in either educt or product is usually split. The system hence splits the molecule and only assigns carbon to both half-reaction, which does not add up. The correct representation of the molecule is also faulty as it was split incorrectly. As such equations only occur rarely and several adjustments would be needed to overcome said shortcomings, I decided not to implement these adjustments, but rather give an error message indicating that such reactions cannot be solved with the system. This is sensible as such reactions only occur rarely and are hardly ever taught in school.

As mentioned before, some other reaction types had to be excluded, namely those requiring the splitting of a molecule consisting of more than two atoms. The splitting of such molecules cannot be decided based on logical rules but requires consideration and an understanding of the underlying chemical structure. For example, the equation

$$KNO_3 \rightarrow KNO_2 + O_2 \tag{5.27}$$

cannot be solved by the system for it is unclear whether the potassium atom is part of the oxidation or the reduction. This only becomes apparent when considering the structural formula which is not part of this system and somewhat contradicts the idea of firstly solving redox reactions simply by following logical rules. On the other hand, including the structural level is a useful extension in the future to further encourage a deeper chemical understanding.

Two more equations had to be excluded to reduce the range of equations to those allowing to build (sensible) half-reactions. For example, if oxidation and reduction basically use the exact same chemical structures, it is impossible to divide these up into oxidation and reduction (personal conversation with ChEx, 18.04.24). Therefore, the equations in lines 34 and 72 of the Excel file (see Appendix D in the electronic supplementary material) are no longer considered. Moreover, some adjustments were necessary to correctly solve the equation where an element that is part of the reduction or oxidation appears twice either in the educt or in the product. For example, when sodium and water react to sodium hydroxide and hydrogen, the hydrogen atom is reduced and appears twice in the product. In such cases, the system failed for two reasons: Firstly, it did not know which of the H-atoms are part of the reduction as this information was only stored within the `find_oxidation_number/6`-predicate and not given onto the `make_halfreaction/5`-predicate. The latter simply took all occurrences of hydrogen in the product to be part of the reduction, which is incorrect. Given our example, only the hydrogen atom, not the sodium hydroxide, is part of the reduction. Secondly, more balancing of atoms was needed. Compare this example after combining the half-reactions:

$$2Na + H_2O \rightarrow 2NaOH + H_2. \tag{5.28}$$

This, however, still is missing a balance of oxygen and hydrogen atoms. Only by adding a water molecule, the formula becomes balanced:

$$2Na + 2H_2O \rightarrow 2NaOH_2 + H_2. \tag{5.29}$$

As there are many such examples and they provide excellent analogies, such examples cannot be left out, so the system has to be able to solve them. However, it proved quite complex to implement this properly, so several steps had to be taken to overcome all difficulties:

1. Firstly, it is unclear whether a molecule such as $NaOH$ is to be split. This far, the rule held that any molecule consisting of more than two atoms cannot be dealt with by the system. Again, a new solution is needed here, as these equations are so crucial. The expert (18.04.24) suggested always putting the element that is not part of the redox reaction to the atom of the molecule that occurs together once more within the equation. This is a common chemical principle, so it makes sense to implement it as a rule. But now what does this mean? Consider the example above: The oxygen is not part of the redox reaction as hydrogen is part of the reduction and sodium is part of the oxidation. So we look for another element that is part of the large molecule but also occurs with oxygen elsewhere. This is true only for hydrogen as hydrogen and oxygen also occur in the educt to form a water molecule. Hence, the molecule is split such that sodium stands alone, but oxygen and hydrogen are grouped together. Note that this also eases the balancing of atoms as oxygen is now always grouped with hydrogen. A new literal of the `make_halfreaction_helper/6`-predicate was needed and the whole predicate was broadened to now include a counter as well as both "OxList" and "RedoxList" from the first sub-step (assigning the oxidation numbers). The counter returns the number of occurrences of the oxidation or reduction element in the equation. This way, if the counter is larger than two, the special case we are discussing here takes place. The new literal then checks if there is a member of "RedoxList" with the attribute no_redox that is part of a large molecule and also part of another molecule in the formula. If so, the corresponding oxidation numbers for the two elements occurring together (that is, oxygen and hydrogen in the example above), are added to know the charge of the resulting split molecule which is then added to the corresponding list. In our example, oxygen has an oxidation number of -2 and hydrogen of $+1$, so the resulting OH^- molecule is added to the reduction. The other element (sodium) is then added as an atom to the oxidation.
2. This resulted in a new split value, namely $Split = 4$. The idea is to mark such equations explicitly using a separate split value. This firstly ensures that the procedure is not applied to other complex equations that need a different splitting. Secondly, the values also imply that the half-reactions need to be put together differently at the end. As the counter is now returned to the `make_reduction/7` and `make_oxidation/7`-predicates (these have also been broadened), a new

predicate called `check_count/5` is added: It takes the oxidation or reduction list, the count, and the current split values as input to return the new split value and an improved oxidation or reduction list. If the counter is larger than two, the new split value becomes 4, and an additional "+" is added to the oxidation or reduction list as it entails one more element than expected. Now the main predicate `solving_redox/2` is also extended to include those cases where the split value changed from 2 to 4, indicating that the product has to be split differently.[4] The if-else-notation in Prolog is used here: If the new split value equals 4, the next step (namely `balance_atoms/4`) is called with a split value of 4, otherwise, it is called with the "original" split value). The same notation is used to call the last step, `make_fullreaction/6`. These two predicates hence need further amendments.

3. To balance the atoms of such equations, a new condition was added, namely the condition that the split value is 4. In this case, the helper predicate `make_balanced/4` is called which receives as input the current formula as well as its educt and product and returns the formula with balanced atoms. Using the `findall/3`-predicate calling `find_atom/3`, all atoms occurring in the formula are found. Electrons and any duplicate elements are then deleted from this list using corresponding helper predicates. Another helper is then called before appending the returned educt and product together to return the balanced equation. The helper predicate, `make_balanced_helper/5`, iterates over the list of elements until it is empty. For each atom in the list, the number of that atom in educt and product is found using list iteration and `calculate_atoms/4`. This atom is then balanced before the next atom in the list is called. The balancing works via the predicate `balance_element/7` which takes as input the educt, product, the number of the element in educt and product, and the element itself. It then returns the new, balanced, educt and product. If the number in the educt and product match, no balancing is necessary. Otherwise, if there are fewer atoms in the educt, all occurrences of the element in the educt are updated using the corresponding multiplier. The same happens if there are fewer atoms in the product: The predicate `update_element/4` updates a certain element of the product by multiplying the coefficient with a given multiplier. This way, after iterating over all atoms in the formula, the formula is balanced out.
4. Lastly, the molecule has to be put together correctly. To do so, a new literal to `add_input_molecule/6` is added to include the case where the split value

[4] Note that, in theory, the split value could also change from 1 to, say, 5, to indicate a complex splitting in the educt. However, as such a case did not occur in the vast number of equations provided, the case is left out here for simplicity. It may be a useful expansion in the future though (see Chap. 6).

is 4. It calls replace_with_molecule/3 which returns the new product with the correct molecule. This predicate now includes cases where three-figured molecules occur first in the list. In this case, all three atoms are searched for, and their coefficients and index values are stored to define the new coefficient and index of the output molecule before their occurrence is deleted from the previous product. Importantly, a new predicate delete_first/3 was needed here to ensure only the first occurrence of an element in a list is deleted as at least one of the atoms (hydrogen, in our example) occurs more than once in the product.

Upon testing it turned out that this complicated add-on to the system pays off: All other equations can still be solved correctly as they have a different split value and are thereby not affected by these changes, but the system can now successfully solve cases where an element occurs twice in educt or product as well.

5.4.5 Modularization

The system is supposed to work in a step-based manner. Each of the steps described above is one important step towards the goal and analogies can be given after each step. This requires great flexibility or modularity of the system. I therefore made the system modular by making those files a module where only the main predicate is needed—that is, all sub-steps. I used Prolog's built-in module/2-predicate which takes the name of the module (or file) and the list of public predicates as input. Such modules can then be called (cf. in *main.pl* or *solving_redox.pl*) via the use_module/1-predicate. This means a predicate of a file is assessable from another file if the first file is made a module and the second file calls that module. However, only those predicates listed in the module declaration are publicly accessible, the other remain private within their file. This way, predicates can be called elsewhere without making the whole code public, which may lead to double occurrences of predicates of the same name or other inconveniences. For example, the *solving_redox*-file obviously needs to access the oxidation numbers to call find_oxidation_number/7. To do so, the use_module/2-predicate is added to the file to call *oxidation_number.pl*. This file is made a module by declaring it so using the module/2-predicate at the beginning of the file. In this case, only the find_oxidation_number/7-predicate is needed elsewhere, so only one predicate is made public, namely module(oxidation_number, [find_oxidation_number/7]). Adding such module declarations to all helper and substep-files allowed me to split the code into sensible files, for example, aggregating all balancing predicates into one file and all predicates to make the half-reactions into another.

5.4.6 Algorithmic Debugging

This concludes the expert module. I will now describe the implementation of the student module using AD to detect errors or misconceptions and the feedback module choosing a promising analogy.

As already mentioned, I would like to use analogies instead of an error library as a feedback strategy here. If a student solves a step incorrectly, the system should register this error and present a corresponding analogous, correctly solved example.

The original idea was to orientate me strongly on the already implemented ITS *Subkraki*, which is an analogy-based tutoring system for written subtraction, both for AD and for the generation of analog examples. Unfortunately, I came to the conclusion that the implementation of AD and the analogies in *Subkraki* are very difficult to transfer to my use case since *Subkraki* proceeds column-wise: With written subtraction, it makes sense to check column by column whether, for example, the correct result is obtained for the one's digit, i.e. whether the subtrahend has been correctly subtracted from the minuend. With chemical formulas, on the other hand, it is completely unclear what to compare here as equations are linear, not in columns. For example, the elements from the reactant and product cannot simply be written one below the other, as it is not clear which are part of the oxidation, and which are part of the reduction. Hence, the *Subkraki*-templates did not help me in my approach. Instead, I think it makes more sense to have students carry out each sub-step individually (at least in the case of an incorrect answer) and compare the result of each sub-step with the result of the expert module (*Subkraki*, on the other hand, always compares entire answers). This procedure trains the rule-based approach to solving redox reactions and an analog example can be generated specifically for a sub-step. As the principle of AD is now slightly modified, this approach represents a deviation from the originally planned procedure. But it makes more sense for the use case, corresponds to chemical guidelines (email with ChEx, 25.03.24), and makes it easier to create analogous examples. In addition, such a system is step-based—in contrast to *Subkraki* or the approach of Zinn (2014)—and meta-analyses have shown that sub-step-based systems are actually superior (VanLehn, 2011).

A further question is whether this step-by-step approach is required in every case. This is not only a question of implementation, but also of theoretical conception. After all, from an educational point of view, it makes a significant difference whether students are always guided step by step to the correct solution or only in the event of an incorrect answer. If you force step-by-step answers, you ensure that students actually go through all sub-steps correctly, i.e. that they do not give the correct answer by chance. On the other hand, students should (at least at some point) be able to abstract from the structure of a single formula. To some extent, this also involves abstracting from the clearly formulated rules and sub-steps towards an intuitive—but still rule-based—understanding of the basic procedure. In this respect, predefined sub-steps can take away a learning opportunity. However, as the ITS is aimed at beginners, and many students have difficulties with the principles behind redox reactions, and the choice of formulas was made accordingly, a step-by-step procedure is implemented here. Achieving a later level of abstraction, such as allowing students to skip steps they know already, is very useful, but no longer part of this work (see Chap. 6).

What does this mean for AD? Just as solving the formula is divided into sub-steps, so is debugging. Accordingly, the "entire" solution of the student is not compared with the solution of the expert module (in contrast to *Subkraki*), but rather the solutions of the individual steps. However, this still corresponds to the principle of AD: It is called individually for each sub-step. It iterates over the solution of the student and expert module until an irreducible disagreement is found. This is identified as an error and forwarded to the feedback module so that an analog example can be generated. Note that in this case no tracing is involved, and any deviance is considered an irreducible disagreement as "small" errors (cf., spelling mistakes) are handled in Python, and it is not necessary to trace to the root of the problem (the irreducible disagreement) because the step is already identified by the procedure (see Chap. 6).

As a next step, the implementation of these ideas had to be considered. Since AD works somewhat like the expert module and constantly calls the predicates of the expert module, it makes sense to implement it in Prolog as well. In addition, my AD is part of the backend, so the entire backend—consisting of the expert module and the student module—is implemented in Prolog. However, I decided against using Prolog for the creation of the frontend as Prolog is not designed for web development. Consequently, I implemented the view as well as anything concerned with processing data structures from the back- onto the frontend in Python. Regarding the simple AD I used, no tracing was necessary as it only comprised the comparison of student and expert module answers. Recall that the algorithmic debugging was adjusted this way to better fit the chemical requirements. The predicate `algorithmic_debugging/5` simply takes the current step (e.g., balancing electrons), the equation, its complexity, and the student solution to the given step, and returns whether algorithmic debugging found an error or not. That is, if the expert solution equals the student solution, the predicate returns "*yes*", otherwise "*no*". The expert answer is found using the `find_expert_solution/4`-predicate, which somewhat resembles the oracle in the original implementation of algorithmic debugging. This predicate iterates over all steps of `solving_redox/2` up to the step it receives as input and returns the answer of `solving_redox/2` for that step. For example, given the step *balance_electrons_oxidation* as well as the equation in question, all steps leading up to balancing the electrons are called: `find_oxidation_number/7`, `make_oxidation/8`, `make_reduction/7`, `balance_atoms/4`, and finally `balance_electrons/4`. The return value (in this case, the equation balanced for electrons) is returned as the expert answer to the current step.

Additionally, it is perfectly possible to iterate over both answers as they are only lists and check where they differ. This is a first step towards a possible AD in the future and its first predicate is already implemented: It iterates over two answers (as they are just lists) and checks if and where they differ (see Appendix B in the electronic supplementary material). This is straightforward in Prolog and can be applied to all sub-steps of the procedure as two answer lists can be compared in each step. This new predicate `same_list/5` takes the correct list (from the expert module) and the list to be tested (the student's answer) as well as a start number as input and returns the list index where the error occurred plus the correct solution to the error—that is, the list element of the correct list at the position with the error. The input number usually is zero as we start tracing the lists from their first arguments. If both lists are empty, so both lists could be fully iterated without producing an error, "*true*" is returned. If the tested list is longer, the empty list is returned along with the highest list index to indicate that the correct list is outrun. If the correct list is longer, the last element of the correct list is returned with a list index out of reach to indicate that the tested list has more elements. Otherwise, the predicate iterates over both lists, checks if their first elements match, and updates the index number. If a disagreement is found, the predicate returns the current index number along with the head element of the correct list.

To compare these lists correctly, though, a correct ordering of the oxidation lists has to be ensured. That is, both oxidation lists should list the elements in the order of their appearance within the formula, starting with the leftmost element. In the case of the student's answer, the frontend has to ensure this ordering. It seems slightly superfluous as humans tend to automatically assume this reading order. The expert module, however, sometimes assigns the oxidation number of a later atom within a molecule first if the oxidation number of the first atom within the molecule cannot be assigned easily. As a result, the ordering of molecules may differ from the reading order. Hence, an additional Prolog-predicate within `find_oxidation_number/7` is needed, namely `order_molecule/3`. It takes the formula as well as the first draft of the oxidation list as input and returns the ordered oxidation list. The predicate iterates over both the formula and the given oxidation list to check if the elements occur in the same order in both lists. If a molecule is found, the helper predicate `correct_order/3` is called with the molecule of the equation and the corresponding molecule of the oxidation list. It iterates over both molecule lists and checks if the first atom of the input molecule is a member of the oxidation list (which it should be), and, if so, puts the occurrence of the atom along with its assigned oxidation number as the new head of the oxidation list. This updated, ordered oxidation list of the molecule is then returned. Upon calling all test suites (see Chap. 6), it was ensured that the expert module still works properly after adding these additional predicates.

5.4.7 Analogies

As discussed, solving redox reactions is a step-based procedure. Once certain expertise is achieved, these steps may vary or overlap, but especially in the beginning, it is crucial to follow the steps carefully to not only get the correct result but also grasp the underlying principle. Hence, not only the expert module is divided into steps, but also the analogies should be presented step by step. This way, in case of an error, the result of each sub-step can be compared with an analogous example, so that (hopefully) the proceedings of each sub-step are understood. Let me now explain how I implemented this.

Firstly, let us consider different errors humans tend to make more closely: Some errors result from inattentive reading or overseeing instructions. These will not be considered further, as an ITS aims to overcome systematic, re-occurring errors, resulting from a faulty understanding of the issue at hand. Systematic errors include errors of omission (forgetting to do something), errors of commission (doing a task incorrectly), and sequence errors (doing subgoals in the wrong order; Zinn, 2011). Adapting this to chemistry, errors of omission may be forgetting one of the sub-steps to take (for example, forgetting to balance the number of electrons), errors of commission mean wrongly executing a sub-step (for example, applying the balancing of atoms to all atoms of the equation), and sequence errors basically implies the ordering of the sub-steps was violated. The student may, for example, first combine the half-reactions into one full reaction before balancing the additional atoms. The ITS aims to overcome all three types of errors. To check the ITS can indeed overcome these errors; test suites are written with wrong answers relating to one of the three error types just discussed. If proper analogical feedback is given to all the test cases within the test suite, it seems plausible to assume the ITS can indeed cope with these typical mistakes (see Chap. 6).

When generating these analog examples, the first question is what constitutes a good analog example, more precisely a good AWOE. After all, chemical formulas are quite different: Does an analog example contain the same substances, the same proportions, the same indices, etc.? Above all, it is important that good analogies have an isomorphic structure. However, it is unclear what such an isomorphic structure can look like in relation to chemical formulae. Nevertheless, there are patterns in chemical formulas or substances that react very similarly with each other and in which the sub-steps are almost identical. All you have to do is replace the name of one substance in the formula with the name of another. Such formulas seem to come very close to the condition of isomorphism: As mathematical formulae are considered structurally similar if they share the same form but are instantiated with different values (for example, $7-4+2$ is structurally similar to $9-5+1$; Zeller and Schmid, 2017), it seems only logical that chemical formulae fulfill the requirement if the same basic structure, the same "template", is filled with different substances. One example is various metals reacting with oxygen. The equation for aluminum and oxygen to form aluminum oxide is

$$4Al + 3O_2 \rightarrow 2Al_2O_3 \tag{5.30}$$

whereas the reaction of iron and oxygen to form iron oxide is

$$4Fe + 3O_2 \rightarrow 2Fe_2O_3. \tag{5.31}$$

Not only are the formulas identical except for the symbols of iron and aluminum, but the sub-steps are also exactly the same, so they are structurally similar, they have an isomorphic structure.

Ideally, the system therefore tries to present such a formula as an analog example. It is also important not to generate the analog examples randomly (again, not in accordance with *Subkraki*), as this can very easily result in incorrect or at least very improbable formulas. After all, not every chemical element reacts with every other. Accordingly, a knowledge base, here in the form of a database, in which various chemical formulas are listed, is needed. Formulas of the same group and the same external form then have a sufficiently similar structure to be able to act as analogs for each other (cf. pattern matching). These examples are presented accordingly in the event of an error. After consultation, I decided to present different analogous tasks if there are multiple errors in one step to further facilitate students' understanding (email with ChEx, 25.03.24). The corresponding "database" as a list of all equations of a certain substance group is provided in the test suite *analogies_tests.pl* (see Chap. 6) and is adapted from an overview of typical chemical reactions in schoolbooks (see Appendix D in the electronic supplementary material).

Hence, predicates calling and using these groups of equations are needed. The main predicate of providing analogical examples is called `check_analogy/6`. It takes the current sub-step, the redox reaction and its complexity, and the student solution to return both the analogical equation and the solution to the current step of the analogical equation. This way, both the equation and the analogical solution can be presented to the student. To do so, it first calls `algorithmic_debugging/5` and checks if the return value equals "*no*". If so, an error in the student solution was detected, so it is necessary to provide an analogy. The helper predicate `present_analogy/5` then takes both step and equation to derive an analogy. It firstly finds a suitable analogical equation by calling `find_analogy/2` and then uses this analogy as input to the previously discussed predicate `find_expert_solution/4` which calls the steps of `solving_redox/2` to derive the solution of the expert module to a given equation (in this case, the analogy) in a given step (in this case, the step where the student's error occurred). The first predicate, `find_analogy/2`, takes the tuple of the equation and its complexity as input. It then returns another equation of the same group as the analogy by checking for the group membership of the input equation and calling the helper predicate `is_analogous_to/3` to return another member of the same group that is not the input equation. This other group member is returned as the analogy that is then, along with the correct solution to the current step, returned and presented to the student.

On the other hand, if the return value is "*yes*", indicating that the student solution is correct, `check_analogy/6` returns an empty list to indicate that no analogy is needed. In the view, this is translated to mean that the solution is correct, so we can move on to the next step.

5.4.8 The Controller to Communicate between Frontend and Backend

The controller part of my Master thesis was implemented using Python and is within the server as both the backend and controller run on the server. Hence, an interface between Prolog (the model) and Python was needed. I used pyswip for that part.[5] It allows the creation of an instance of the Prolog class and then uses this instance to

[5] Pyswip can be downloaded here: https://pypi.org/project/pyswip/0.2.2/.

consult Prolog files and make queries to these consulted files. The controller consults all files needed for creating analogies and conducting algorithmic debugging to get the output of these queries in Python. It returns a list which is either empty (if no error was found, so no analogy needs to be presented) or contains the deduced analogical example. However, the output of such a pyswip-based list actually is a dictionary and the quadruples of elements and molecules were considered as separate entries to that dictionary. A single magnesium-atom is depicted as "$', (mg, , (1, , (1, 0)))'$", and an analogy looks as follows:

$$
\begin{aligned}
&['A' : [', (cu, , (1, , (1, 0)))',' plus',', (o, , (1, , (2, 0)))',' arrow', \\
&\quad [', (cu, , (1, , (1, 0)))',', (o, , (1, , (1, 0)))']],' As' : [', (cu, 0)',', (o, 0)', \\
&\quad ', (o, -2)',', (cu, 2)']]. \qquad (5.32)
\end{aligned}
$$

Hence, some parsing was needed to "translate" the pyswip output into the expected format of $(mg, 1, 1, 0)$. This was done by first converting the output into a string and removing the outmost square- and curly brackets as well as the A and colon up front. Next, the `.replace()`-function was used to get rid of all quotes, all spaces, all double commas, all colons, and all brackets with commas. The first method of the parsing class returns the result of this first parsing. Now the output has to be separated as both the analogy and the solution of the current step are returned as one by pyswip. To do so, the second method splits the string before and after the Analogical Step denoted As, indicating the start of the step's solution. The corresponding parts are stored in two variables that are then returned. Lastly, only for the analogy, another step is needed to remove the inner parentheses between the quadruples representing atoms or molecules. Regular expressions are used here to ensure that only those parentheses directly enclosing digits are removed. That is, all parentheses directly before or directly after a digit, but no digit opening or closing an atom, are removed. Both the analogy and the solution to the current step of the analogy are then returned.

Other than that, the Python part of my thesis included some more modules following a simple process, all aiming to successfully communicate between frontend and backend. This is part of the *app.py* file. The process works as follows: The file opens the Flask App and starts the homepage. It then chooses an easy or a difficult equation and its complexity from the list of all equations. Now the main part of the Python part is to ensure a continuous user interaction: It has to iterate over all steps to solve a redox reaction, present each step to the user, take the input of the user, present it to the expert module, receives the answer and either depicts an analogy if the student answer is incorrect or moves on to the next step if the correct answer was given until all steps are solved (see Fig. 5.3).

```
Input : List of further_steps, all steps to solving redox reactions
app ← start_flask_app
while true do
    app ← show_homepage_template
    equation, complexity ← random.select(equation)
    i ← 0
    while still steps in further_steps do
        step ← further_steps[i]
        move_to_next_step ← False
        while not move_to_next_step do
            app ← present_step(step)
            expert_module ← get_student_answer(answer)
            parse_output(parsedA, parsedStep) ←
              expert_module_answer(analogy,step)
            if wrong answer then
                app ← present_analogy(parsedA,parsedStep)
                continue
            else
                move_to_next_step ← True
            end
        end
        i ← i + 1
    end
    app ← show_solved_template
end
```

Fig. 5.3 Algorithm Depicting Communication Between Back-and Frontend

5.4.9 The Frontend

The frontend builds the fourth and final module of an ITS and allows the user to interact with the actual system. In this case, the frontend is only a prototype as the focus was on ensuring the expert module works flawlessly. However, based on the conducted interviews, it is straightforward to use and has an appealing design. Furthermore, following Almasri et al. (2019) distinction, I have implemented a component interface.

I built this fourth module using Flask, a JavaScript-based engine for front-end development in Python.[6] The application for this thesis runs on a single local device. If called, the app is started and configured, so the frontend can be used. The implementation follows a simple server-and-client architecture: The server—as part of the backend—controls the application and redirects it to the desired pages. It returns html-templates which are then presented to the user. These templates constitute the actual frontend. The server keeps track of which step the user is currently solving, ensures the correct equation is depicted, and calls the backend module with the user input. However, the parsing from the user input into Prolog syntax is still missing, so the user input has to be readable by Prolog. More precisely, the helper class *pipeline.py* receives the user input (a Prolog list) from the server and ensures the answer to this input from the expert module is depicted correctly by calling the parsing class, which parses Prolog output into a more readable string, described above. The server then receives this parsed answer from the backend and either instructs the frontend to show a helpful analogy or, if the answer is correct, it moves to the next step and asks the frontend to show it to the user. Furthermore, after consulting with ChEx, the user can choose between easy and more demanding equations, so both can be practiced.

Regarding the frontend, the main template called *base.html* defines the basic structure of the entire application. All other templates inherit this information and then build up the necessary containers for the user interaction. I hence follow the template inheritance affordance of Flask. There are seven other templates, one for each type of user interaction to fulfill the nested loop described in the previous chapter: First, the homepage (*index.html*) depicts several options, including choosing an equation or receiving more information either on redox reactions or how this tutoring system works (see Fig. 5.4). The corresponding templates *information.html* and *tutorial.html* can be accessed correctly, but do not yet provide much information, so they need to be enlarged later. To solve the redox reaction, templates exist depicting the current step and asking the users for their solution, a template depicting that the previous step was answered correctly, another template depicting an analogy in case of an error (see Fig. 5.5), and lastly a closing page that redirects back to the homepage if the last step was solved successfully. Additionally, to make the frontend more user-friendly and appealing, I used bootstrapping.[7]

Several issues occurred when building the frontend which will be mentioned here. Firstly, some minimal error handling was added to ensure the backend can always be called correctly. All cases are covered by the backend module, it only

---

[6] Find the documentation on Flask here: https://flask.palletsprojects.com/en/3.0.x/.

[7] Find the documentation of bootstrapping with its stylesheet here: https://getbootstrap.com/.

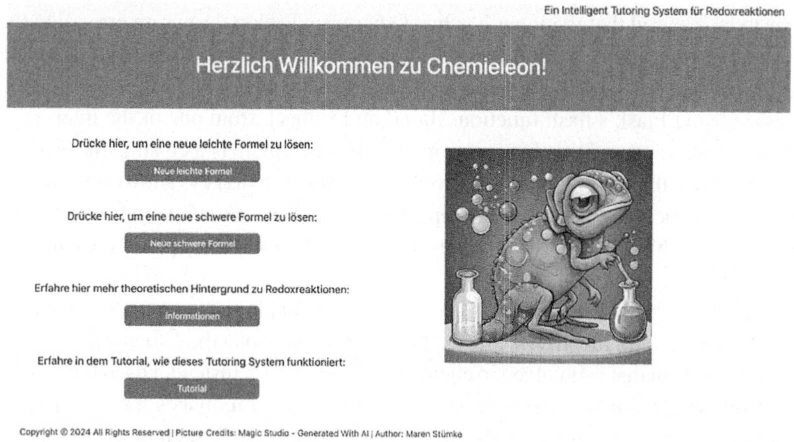

Fig. 5.4 Homepage of the implemented frontend. Note that the depicted *Chemieleon* was created using an AI by Magic Studio (28.05.24)

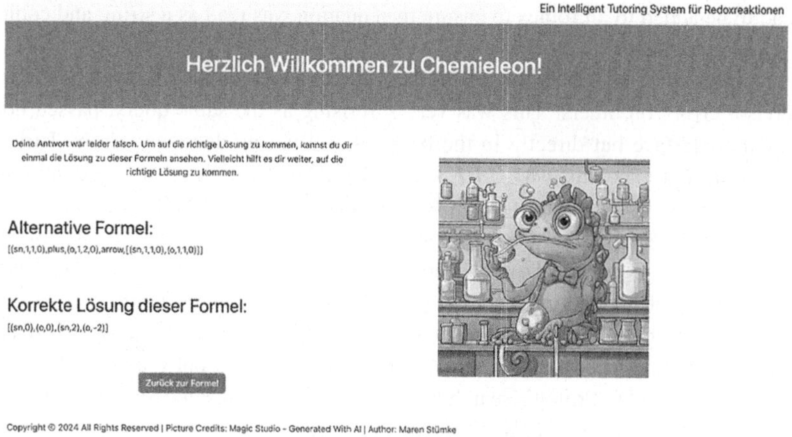

Fig. 5.5 Incorrect answer template of the frontend. Next to the *Chemieleon* (created with AI on 28.05.24), an analog equation and its solution are depicted

has to be ensured that the input is a list. Otherwise, Prolog throws an error, and the system is interrupted. Therefore, a script is added to the *next_step.html*-template to ensure the first typed character is "[" and the last one is "]" to ensure a list is passed using Flask's flash function. Based on feedback from one of the interviews it is now also ensured that for every opening bracket, there is a closing one as well.

Secondly, the server includes a dictionary with all steps as keys and a user-friendly notion as values. For example, the step *find_oxidation_number* translates to "die Oxidationszahlen". Only these values, the description of the step, are presented to the user in the frontend.

Thirdly, it proved difficult to find a publicly available picture suitable as a logo for the application. Ideally, of course, this logo should depict the *Chemieleon* itself, so a chameleon that is involved in chemistry (see Fig. 5.4). Instead, I used the picture AI from Magic Studio, an online website, to create several images for the different templates of the app and declared this usage in the copyright section in the footer.

Fourthly, the passing of parameters onto the HTML templates was more difficult than expected as the equation (a list of quadruples) was not read as a list, but rather as a nested, complex dictionary that could not be shown to the user. Type declarations had to be added to all routes to ensure the equation was read as a string and could be passed correctly.

Lastly, some of the sub-steps worked fine, whereas Prolog returned a "zero divisor error" on others. This was very surprising as the same query, passed not via the interface but directly in the backend, did not produce any errors. I tried to ensure the encoding was *UTF8*, traced the errors, updated my pyswip version, and briefly switched to swiplserver, another Python-Prolog interface, but the errors persisted. I eventually solved the issue by adding more cuts to the `find_expert_solution/4`-predicate in the backend. This solved the issue, yet I could not find any information to work out a better understanding of why it solved the problem. I suspect the pruning did not work correctly with the interface because the interface does not directly query the backend, but rather works as a new, higher level to Prolog's search tree, so more cuts are needed to prevent wrong backtracking which (probably) resulted in the errors.

Once these issues were resolved, I tested the frontend via iterating over all steps both for the easy and the difficult equations, and checked that all redirecting and the presentation of the analogy worked correctly. Further testing was done during the interviews (see Chap. 6).

6 Realisation and Evaluation

Firstly, I will lay out some technical details of my implementation. Full details on all technical aspects and the libraries used are provided in the appendix (see Appendix A in the electronic supplementary material). A MacBook Pro with an Apple M1 Chip and 16 GB memory is used with the operating system macOS Sonoma in version 14.5. Furthermore, the programming languages and software run on the following versions: Python version 3.8.9, Flask version 3.0.3, pyswip version 0.2.11, and SWI-Prolog version 9.2.2 for arm64-darwin. Now, in this chapter, the ITS is tested followed by a short qualitative evaluation in the form of an interview analysis with two people. This ensures that the ITS works properly and has accessible usability. Next, the chapter includes a discussion of the findings, including limitations as well as advantages of the presented approach compared to other systems. Lastly, the chapter closes with an outlook on possible future enhancements to the system.

6.1 Testing

Generally speaking, software testing aims to find symptoms of erroneous behavior. However, such testing is never perfect as it only reveals the presence, not the absence, of errors. Testing is executed via planning, preparation, execution, and evaluation. Hence, I use two test suites consisting of several test cases both for the background module and the analogy, so the input and predicted output are tested in both cases. As I consider not all possible paths within the program—given the large number of predicates this would be too complex—the test suites cover black-box, but not

---

Supplementary Information The online version contains supplementary material available at https://doi.org/10.1007/978-3-658-51032-9_6.

M. Stümke, *Identifying and Overcoming Student's Misconceptions in Chemistry*, BestMasters, https://doi.org/10.1007/978-3-658-51032-9_6

white-box, testing. These are functional tests focusing on input-output behavior. The input representation is chosen in such a way that all output behavior is exhibited. This usually includes examples of equivalence partitions and class boundaries, so normal as well as corner cases. White-box tests, on the other hand, are structural tests with a set of inputs so all program paths are covered.

To test both the expert module and analogies, 3–10 typical error cases per analogy and input formula are given to account for background testing. The test files ensure that if true is returned, all queries are answered correctly. For example, a chemical formula and its correct solution are given as input to `solving_redox/2`. This way, the correct solution is tested against the solution found by the expert module. If they match, true is returned, thereby revealing that the system also solved the equation correctly.

The test suite is provided largely by an Excel file the expert gave to me. It includes many equations grouped by similarity. However, we decided to leave out two complex categories the system cannot cover and that are not usually taught at school. Namely, we dropped the group "Verbindungen" (compounds) and two items of the group "Metall + Wasser" (metal and water). The rest was used both for analogies and the expert module. I rounded off the test suite by adding a couple of equations retrieved from different textbooks on chemistry and the website http://www.leifi-chemie.de.

I applied the test suite to three test files not only aiming at different modules but also with slightly different targets. The layout of these test files is taken from Zeller (2015) but adapted to fit the use case. Firstly, *expert_module_tests.pl* aims to detect bugs within the expert module by (semi-)automated testing. It calls the `solving_redox`-module to test both correct and incorrect cases. The predicate `all_correct/0` lists all correct reactions, such as metal reacting with oxygen or compound reactions, thereby calling the test suite listed below. As expected, this predicate returns true, indicating that the expert module works as intended: No matter which redox reaction is given as input, it returns the correctly solved reaction (or, in this case, it returns true, as the result of the expert module equals the actual solution to the redox reaction). The predicate `all_incorrect/0` lists types of reactions that are false, either because they are not redox reactions, they are disproportionist equations, or they cannot be handled properly. Therefore, it summarizes those edge cases discussed previously in the chapter on the expert module (see Chap. 5). Put together, this test suite supports the claim that the expert module is indeed able to solve (most) redox reactions and correctly identifies faulty input equations.

Secondly, the file *diagnosis_tests.pl* aims to understand if the algorithmic debugging—and, therewith, diagnosing errors—works properly. Thus, it uses the modules of algorithmic debugging along with *solving_redox* to check if the correct misconception or error is discovered given an erroneous input. Once again, correct and incorrect tests are established and listed in the main predicate `all_diagnosis/0`, and the test suite is listed. This time, the test suite is adapted to include examples of typical errors students tend to make when solving redox reactions. I then checked if the correct cases were recognized as correct. This time, however, not all redox reactions are checked but only its substeps. Recall that the algorithmic debugging is implemented stepwise, so an incorrect answer has to be detectable in each step. For example, two assignments of oxidation numbers are compared. If they are identical (and both correct), the system is expected to return true. On the other hand, if one of the assignments (representing the student's answer) contains an error, the algorithmic debugging is expected to find the error and return an irreducible disagreement when assigning oxidation numbers. To do so, the correct cases (two per step) are tested using the `all_correct/0`-predicate via calling `correct_student/0`. It calls the `algorithmic_debugging/5`-predicate with the step, the equation, the "student" answer (which is correct, in this case), the answer given by the expert module, and the expected return value "*yes*". More importantly, typical student errors on each step were tested using the `all_incorrect/0` or `incorrect_student/0`-predicates which again calls the algorithmic debugging with an incorrect student answer and the expected return value "*no*", so the predicate evaluates to true. These errors are based on a discussion on typical student errors on each step with ChEx (mail from 07.05.24). For example, when assigning oxidation numbers, students typically do not take the sum rule of molecules into account or forget to consider indices. They also tend to assume symmetry within molecules, meaning all elements should have the same oxidation number, make errors calculating the electronegativity, or take a coefficient or index to be the oxidation number of a single element. Accordingly, all these cases are covered by examples in the test suite and the comments behind the examples indicate which typical error this example aims to demonstrate. I tested the system with different error types to see whether the system recognizes all typical errors reliably. However, it is not enough to abstractly identify an error type, for example, "Did not consider the sum rule in the molecule", to be able to give concrete examples. For that, I need an example in which exactly this error was made. I can then give this example as input (e.g. an incorrect determination of the oxidation numbers as a list) and if the system recognizes the solution as incorrect, it can be assumed that this error type is correctly identified as an error: In case of the sum rule, not the correct oxidation list

$$[(h, 1), (o, -1)] \tag{6.1}$$

is given as input for the molecule H_2O_2, but

$$[(h, 1), (o, -2)] \tag{6.2}$$

instead—and it was recognized as an error.

Upon calling these predicates, it was shown that the system correctly distinguishes all correct from incorrect answers in each step. Seemingly, the algorithmic debugging works as intended.

Thirdly, a test suite helping to return an analogous example and check if it indeed provides an analogy, namely *analogies_tests.pl*, was created. It works slightly differently from the previous two as the suite does not comprise test cases for analogies, but rather checks if a provided counterexample is indeed an analogy. As all equations within a substance group are considered analogous to one another, it lists all group members of the different groups using `group_member/2`. For example, aluminum and sulfur reacting to aluminum sulfide is part of the group *metal_and_sulfur*. The main predicate `find_analogy/2` takes a formula and returns an analogy by checking which group the formula belongs to and calling the helper predicate `is_analogous_to/3`. The latter receives a formula and the group it belongs to and then finds another member of that group randomly and returns it as the analogy. However, a problem occurred here as there was only one equation of the group *nonmetal_and_halogen*, that is, hydrogen and chlorine reacting. This is an obvious shortcoming as no proper analogy could be presented to a student if an error occurred in this equation. But the expert provided three "new" equations belonging to this group that are now added to the analogy test suite. These equations are good analogies as the procedure of solving them is very similar to the reaction of hydrogen and chlorine.

The test suite now works on all groups when producing analogies, but testing these predicates also ensures the analogies work correctly: To check if a correct analogy is returned for each group, two cases are tested per group. Each time, `find_analogy/2` is called with an equation of the corresponding group. It is then tested if the returned analogy belongs to the same group as the input equation by using the analogy and the group as input to the `group\_member/2` predicate. If a correct analogy is chosen by the system, the test case returns true. Once again, all tests (that is, testing all groups) are added to the `test_all/0`-predicate which can be called to ensure the analogies work correctly. Upon calling this predicate, it returns "true", thereby giving evidence that generating analogies works as intended.

Taken together, these three test suites suggest that the expert module, the diagnosis of errors using algorithmic debugging and generating analogous examples works as intended. The background module is therewith functional and complete.

The Python part (the controller) was not tested thoroughly as it mainly communicates between the backend and frontend which makes testing somewhat superfluous: If both GUI and backend work as intended and no errors occur when they interact, the Python part also works as intended.

6.2 Interview Questions

However, a short evaluative study was needed to test the frontend, more precisely the usability and comprehensibility of the frontend as the implementation of the use case. The results are not generalizable, but they should provide some initial indications of sensible future improvements. And if the interview study is long and detailed, few interviewed people suffice to gain insights. Accordingly, two people tested the implemented (simple) interface qualitatively and extensively, one focusing on the usability, and the other one focusing on the chemical realization. Both answered a couple of interview questions regarding their user experience. The idea is to use the feedback to gain a better insight into sensible future additions to the system. In this respect, the interviews are to be understood in particular as an outside look at future enhancements and extensions. In this chapter, the design of the interview study is briefly explained before, in the next subchapter, the results of the interviews are evaluated.

6.2.1 Interview Concept

The two interviews each took about two hours and consisted of three main parts: First, the format and the procedure were explained, in particular the expected Prolog input. Next, both test subjects were asked to solve one task per category (i.e., an easy and one difficult task). It was originally intended to solve two easy and one difficult equations, but as this procedure proved to be very time-consuming (both needed about 30 minutes per equation), I decided to only let them solve one equation from each category. Notes were taken using the think-aloud technique: While they were solving the tasks or interacting with the frontend, they said what was going through their mind and how they rated the interaction. Third, both answered six open questions, some with follow-up questions, about the application they used. A time of 3–5 minutes to answer each question was calculated which roughly worked out. Some of these questions are general and were asked to both participants, some referred more to the usability or the chemical aspects and were asked to the prospective participant. Then there was an approximately five-minute summary on my part as well as an outlook on how the results of the interview influence the further procedure in the context of this master's thesis. Lastly, it was asked if there were any final questions. I created an interview manual guiding through the steps just explained to ensure a smooth and comparable process. This manual looked fairly similar for both interview partners, only the third part—answering open questions—differs (see Appendix C in the electronic supplementary material).

Overall, the interview manual was based on the TAToo interview standards (Koch and Westhoff, 2012) with some necessary adaptations to the use case. The interviews were conducted in German and the notes are taken in German as well. Regarding the open questions, the guidelines from Thielsch et al. (2012) were taken into account. Consequently, I tried to ask easy-to-understand, short, and open questions with enough time to answer each question thoroughly. Moreover, I sorted the questions from general to specific and avoided double or leading questions as well as double negatives. As both interview partners answered each question with ease and I received the answers I was hoping for, the questions seemed to be sufficiently clearly formulated and easy to understand.

6.2.2 Interview Analysis

As already indicated, only two equations were solved in both interviews, as solving these while testing the system and thinking aloud required a significant amount of time. In addition, both participants said that solving a third formula does not provide further insights. They both agreed to be named, so I will use their names for the analysis.

In the first interview focusing on usability, the first interviewee $P1$ praised the clean and simple design of the application as well as the color scheme, which is not distracting and makes you feel comfortable. The pictures are considered cute and enhance the application. The homepage is overall attractively designed, but the difficulty of the formula could be chosen via a drop-down menu. Also, dividing the solution into steps was rated as positive: Especially the *correct_answer.html* template was rated positively, as the steps are solved individually: You solve one step, have this intermediate screen as a resting point, and only then the next step follows. One also feels accomplished and receives confirmation, especially after a potentially frustrating prior step.

The forwarding worked smoothly and was quite intuitive. The requirements were always clear. In general, the system was very intuitive to use—except for the input format. However, $P1$ said it posed no real problem for her. The buttons are labeled well so you know what happens if you press them. Together with the simple, beautiful design and the images, the usability of the interface was rated very positively.

In this respect, the formula check in the backend (and thus the backend itself) has to work very well, as—although solving redox reactions is a difficult use case—there were never any problems, and the errors were consistently caught, even when simply clicking through. This allows one to concentrate fully on solving the formulas.

Of course, an improved input format is necessary. It is better if the correct solution of the previous sub-step is displayed when solving the next sub-step, as it is necessary to take notes on paper to remember the previous partial results at the moment. Displaying results eases the handling and prevents slipping through the steps or getting them mixed up.

She deliberately built in a few errors in almost every step to check the robustness of the system. For example, she typed in nothing, reloaded the page, mixed up "+" and charge, made syntax errors, and rearranged the correct formula in different ways or typed in wrong numbers. In principle, the system was able to intercept every case (except initially the case where an opening bracket was not closed, which is now fixed), so the system did not run into errors. She generally recommended catching all form errors such as bracket or comma errors directly in the frontend and only calling the backend and analogies for content errors. However, she was unable to deduce the type of error from the feedback, as the same feedback, an analogy, always appears—except in the case of syntax errors, which are intercepted well. The representation of these analogies does not adapt to the type of error. The analogies seem to be very useful and helpful per se, as she immediately recognized the structural similarity of the formulas. For example, she commented that chlorine is now in the formula instead of fluorine, but the structure is the same. As the electron in the analogy is simply negatively charged and the equations look so similar, this is probably her error: The electron seems to depend on hydrogen and this substance is the same in the analogy and the original formula. At the same time, she would not know how to arrive at the correct solution. Individual error handling such as highlighting or a button that explains the relevant step again would be useful here. In addition, the task should be presented next to the analogy, that is, the screen should not only display the analogy, but the erroneous solution as well, to ease the comparison.

Lastly, the footer entailing copyright information, especially when displaying the next step, is too large and therefore irritating. The header could also be smaller. Instead, the specific instructions and the next step should be highlighted. A button redirecting to the information page is also useful if the answer is correct and you want to read what comes next before proceeding to the next step.

The first possible improvement is to incorporate two modes. One mode works as before by executing all sub-steps one after the other. A second, expert mode could skip some sub-steps or at least list every step below the other on one page and only check if the steps are solved correctly. Furthermore, the theoretical background and the tutorial pages (both promising ideas) should be supplemented, and it should be possible to be redirected to these explanations from certain steps. The analogies per se should be retained as a feedback mechanism, but specific markings or individual error handling, such as highlighting or small explanatory texts, should be added. Especially if you have little background knowledge of redox reactions, it is otherwise very difficult to link the error and the analogy together. At the same time, the design, the analogies themselves, and the feedback on and between the steps should be retained.

The second interview focused more on the chemical principles. This was also evident in the way the participant *P*2 interacted with the system. He made fewer deliberate input-related mistakes but tried to understand the chemical structure of the system and the reactions depicted. For example, he expected the formula to remain the same throughout the steps, which is the case. The input format was a major hurdle here, as the formula representation is very unfamiliar and *P*2 therefore considered it to be unsuitable for students, especially since they would then have to focus on the requirements of the format, not on solving the task. Some of the sub-steps are not formulated precisely enough. For example, it is unclear what exactly should be done in the "Atomanzahlausgleich der Oxidation" (balancing the atoms of the oxidation). Firstly, the previously determined half-reactions should be displayed for this, and secondly, it is not possible to deduce the desired answer format from the question, as the question asks for balancing, but a formula is required as input.

*P*2 also found it difficult to deduce her own mistake precisely from the analogies. After all, since the equation is not displayed again, students are required to keep the step and their solution in mind. This is said to be not only unnecessarily complicated but also requires a high mental load and is therefore prone to many errors. For example, students could misremember what their previous answer was and therefore misinterpret the analogy. Especially if you are unsure about your solution, you do not remember your answer correctly. Therefore, incorrect answers and correct analogous solutions should be displayed next to each other to ease the comparison.

Furthermore, in the balancing electrons step, the same equation was requested for the third time. Although the balancing makes sense as a test step, students would be annoyed. Accordingly, the answer format should be adapted so that the previous, own solution is displayed, and you can decide whether you want to stick with this answer (that is, whether the electrons in the formula are already balanced) or whether you want to type in a new answer. This shortcut of using the previous input again can be realized via a button: If you press the button, the old formula is sent to the backend. This means only local changes are necessary without affecting the expert module. Otherwise, the content and the ordering of the steps make sense. Only the step "weiterer Atomanzahlausgleich" (further balancing of atoms) is superfluous for easy equations, as this is never necessary. If an easy formula is selected, this step can therefore be omitted. In the case of more difficult equations, the instructions are unclear as to which substances have to be balanced. For example, you could assume that you should also balance substances that are not involved in the redox reaction, but this is not intended by the system. *P*2 hence did not know what was expected as input. The instructions have to be clearer here or better adapted to certain chemical equations.

Overall, both interviewed persons suggested some improvements: The input format should be revised. You should be able to see your input when comparing it with the analogy. It should also be possible to compare the current sub-step with the previous solution and simply take it on. The steps themselves should not be reduced, but the effort required to go through them should be minimized to avoid redundancies or frustration. Some of the tasks' instructions (that is, only the formulations, not the coding or the chemical steps per se) should be sharpened and made more precise to avoid ambiguity during processing. They could also be tailored a little more to the target group of (young) students.

On the other hand, they stated that the analogies are very useful as they encourage students not to simply adopt a displayed solution, but to think through the structure and find their similarities. The examples have a very parallel structure and since the same element constellations are used, it is possible to deduce the error from them. The analogies themselves should therefore not be changed. It also makes sense for the structure of the system to be identical in the case of the easy and difficult formulas to internalize the sub-steps and the underlying concept. Again, it was emphasized that the information on redox reactions and the tutorial are a useful addition. The tutorial should primarily explain how the system works, so it should focus on usability aspects. The different task difficulties make sense, and the steps are paced well. In the event of an error, an immediate hint is given that does not explain the error but offers a comparison. In this respect, the type and structure of the feedback should be retained. The chemical implementation in Prolog also appears to be sensible, insofar as it can be evaluated and works flawlessly. Overall, the tool was rated positively and could be used in lessons. It allows a systematic practice of the steps of redox reactions until they are internalized.

As an extension, it was recommended to further differentiate the examples and to grade or assign them to content groupings. From a teacher's point of view, these groupings could provide insights into how well the students perform, how many students in the class use the tool, which typical problems arise, etc. Furthermore, user modeling could be useful because the system remembers if a person no longer requires certain sub-steps or should be set more difficult exercises. Gamification elements could also increase extrinsic motivation to use the tool. Such playful incentives could be collecting points or rewards. Moreover, if the analogies do not help, additional content-related information or even easier examples could be displayed. Balancing could be supplemented by a symbolic representation on a sub-microscopic level to promote a deeper understanding of the reactions and the actual chemical particles interacting.

6.3 Discussion and Limitations

I will now critically discuss the implementation and its evaluation. Firstly, a reflection on some chemical issues is useful. As mentioned before, not all equations could be considered. Some of these exclusions are not that critical as they concern chemical edge cases that are hardly ever used in schools and never to teach redox reactions in the first place. An example of this is a disproportionist reaction. There are some other problematic categories never taught at school that are left out here, such as formulas having multiple solutions as well as formulas having no solution at all.

Moreover, most molecules with more than two atoms are not considered as it is unclear how to split them into half-reactions. This is a slightly graver exclusion as quite a few molecules contain more than two atoms. However, once again, such molecules tend to be part of very complex equations not used as examples in school. The problem is that their splitting does not follow logical guidelines. It is not rule-based, it depends on too many chemical factors to make general rules. Hence, even if it is possible to code the splitting of large molecules in Prolog, it somewhat contradicts the idea of logic programming. One would mainly code error cases or exceptions, not rules. It may be easier to code this part in Python and put the output into the Prolog code. It should be noted that although the system cannot solve such equations, it identifies the error correctly.

Lastly, the system this far solves redox reactions correctly but the backend cannot yet solve word problems (Textaufgaben) although they are frequently used in chemistry lessons. This is a true system limitation and should be included in the future (see below).

Next, I consider the implementation of some chemical concepts with inherent limitations (that are not necessarily shortcomings): The solution to at least one equation is not fully shortened as there is no extra predicate ensuring the shortening of equations. This, again, is an edge case as it is hardly ever necessary to shorten a chemical equation. Moreover, an unabridged equation is not mathematically wrong, just imprecise. An additional test step to prove the correct balancing of all atoms is not implemented either as it hardly seemed necessary. It does not make sense to check if a computer correctly carried out an arithmetic operation for a computer—unlike students—does not make calculation errors. It is hence more of a requirement to the front, rather than the backend, to provide a tabular overview of all atoms involved in the reaction, so students understand the involved elements.

The sorting of an equation always requires electrons to be placed at the end of an equation and additional atoms (such as H_2O) at the beginning of the equation. This is not a chemical rule, though, they could be placed anywhere. But it was necessary to stipulate a fixed ordering here for the predicates to be coded accordingly, so it was clear where to look for electrons, for example. However, even though it is not a rule, it is the most common ordering of a chemistry equation, so students will likely follow it.

Several adjustments were necessary to account for the case of an element appearing twice in the educt or product. A split value of 4 was introduced to solve this issue meaning the element occurs in the product twice. This is incomplete as the element may be appearing in the educt twice. However, this never occurred in the test suite. As the examples in the test suite provide a thorough range of redox reactions, I take it to mean that if something does not occur in any of these equations, it will most likely not be covered in chemistry lessons in schools. I hence believe it sufficient not to implement the corresponding split value of 5 here. On the other hand, the test suite is not exhaustive, so it may be useful to implement it in the future, to further extend the number of equations the system can solve.

Moreover, the handling of recombination equations proved to be very difficult. In such combinations, certain atoms that are neither part of the reduction nor the oxidation occur. The question was how to assign these into half-reactions and whether to balance them. Several ideas were considered here: Leaving such examples out was not an option as they occur frequently in chemistry. Such particles (often oxygen) could be left out in the half-reactions. However, it is unclear how to consider them when producing the full reaction. It might also confuse students if the oxygen suddenly pops up again. Hence, we decided on simply "adding" the atoms to the half-reactions but emphasizing them: The idea is to (eventually) highlight the "interfering" elements in a different color, e.g. grey them out so that they are still visible, but not on an equal footing with the reduced or oxidized elements of the respective reaction. This is a slight shortcoming, as it is only a graphical solution, and the grey color of these elements has to be explained properly in the tutorial of the application. On the other hand, the correct format of the half-reactions is retained but represented in a more differentiated way. From a chemical and a didactic point of view, such differentiation may be a useful distinction as the active elements being reduced or oxidized are separated from the passive elements just occurring in the reaction. Hence, this solution may help students understand the reaction better (mail with ChEx, 21.04.24).

As the last aspect regarding chemistry and the usability of the tool, the representation itself of atoms, molecules, and equations can be considered. The input is a little complicated but, based on the interviews, straightforward to understand. It is mainly difficult to type in, not to read, so the equations in the backend hopefully are understandable with atoms as quadruples, molecules as lists of atoms, and equations as lists of atoms and molecules. However, lists occurring at any point within lists (as molecules can occur anywhere in an equation) made debugging almost impossible (see below). Nevertheless, I chose a convenient representation as it allowed one to always assign the charge and the number of elements correctly and unambiguously. Moreover, as Prolog is list-based, working with lists intensively in the backend greatly simplified the processing.

Secondly, let us consider the use of analogies carefully. Students perform each sub-step of solving redox reactions, their answers are compared, and an analogy is presented in case of a wrong answer. However, the original idea was to let the system generate analogies based on a tracing of the input equation, so tracing could be used to generate better analogies as they are built specifically around the error made by the student. For example, if the number of electrons was calculated incorrectly, an analogous formula could be built around this wrong part of the input formula. This attempt failed since such an approach is prone to errors as many combinations of elements to create equations do not exist or do not make any sense chemically. One would need many rules and libraries on which elements can be combined chemically given specific circumstances. This is both computationally expensive and very complicated, and it is probably only possible if the scope of your system is limited to include certain substances only. For example, taking only a few chemical substances that react relatively predictably with certain other elements such as oxygen or halogen, and building equations out of these might work. However, it comes at the great price of only covering a tiny part of the spectrum of chemical reactions. Note also that such created analogies are often not as unambiguous as they seem, so analogies created by the system may allow several solutions (see Footnote 3 in Chap. 4). Instead, a tracer can be used to see which rules are called in the input equation and create analogies following these rules. But this is basically what I implemented, as the equations always follow the same steps.[1] There are some more complex examples where additional rules apply (mainly additional balancing). Perhaps a tracer could be built specifically for these examples where it recognizes the

[1] To be fair, some of the steps are not always necessary, but they are still called to ensure students internalize all steps needed to solve redox reactions.

faulty step and generates an analogous example based on the previous step but with the correct solution to the current step. Such an approach would follow strict rules and allow the creation of new equations, making the system more generalizable. But, again, building a step out of a previous step can easily lead to incorrect formulas if some chemically unstable compounds are created. It would have to be ensured that the equation generated by the system is correct, e.g., by testing it against a database of possible chemical equations and regenerating the equation until it occurs in the database. Put together, if formulas are self-generated, many restrictions apply and the chemical spectrum is not covered.

At this point, it is hopefully clear why I chose my approach and discarded the original idea: Why go through the effort of coding a system capable of building equations if, in the end, the equations have to match given equations in a database? It is much simpler, much more efficient, and less error-prone to choose a fitting example from the given database as the analog. Furthermore, this approach takes more elements and special cases—those listed in the database—into account, and the analogy is always correct (given that the database is correct, of course). Note that this is not the same approach as error libraries (see Chap. 3), as the main feedback mechanisms still are analogies, not error libraries. A proper analogy is simply chosen from a database. The database does not need to provide perfect error feedback or entail all possible errors made, so the approach does not include the disadvantages associated with error libraries.

Alternatively, creating the analogies merely for oxidation numbers was considered. The idea was that the output looks slightly different in this crucial step, so it may be easier to generate it automatically. Yet again I do not believe it works because replacing an element with another in an oxidation list is chemically wrong. Recall that such a list consists of tuples of elements with their oxidation number. The oxidation number depends on the equation and the other atoms or molecules in that equation, so it is not always possible to derive an oxidation number from an element itself. Hence, again, the actual step of assigning oxidation numbers has to be taken to provide a proper analogous example—which is exactly what I implemented.

Instead, structurally analogous equations are provided that are based on the substance groups in chemistry which provide similar equations. Each time, a different analogy from within the same group as the input equation is created. Is this a good choice? On the one hand, if a person makes several mistakes one after the other, that person is probably really struggling. Lots of different examples might then not be helpful, but rather confusing. Hence, it could be considered showing the same analogy per sub-step. On the other hand, if one is presented with several analog examples, one practices to look for their structural alignments which can greatly enhance the understanding. This is why I decided for the second option. However, another helpful option for stuck students is sensible, such as highlighting the decisive factor for the correct solution. If, for example, a student has not balanced the negative charge, the corresponding molecule in the analogous example is marked (see below).

In general, the test suites reveal that the analogies work as intended. They are also rated positively in the interviews and the test suite always returned the intended result, thereby supporting my implementation choice.

Thirdly, I originally intended to adapt the technique of AD to the student module of my ITS and use it for error detection. It was meant to trace over a student's answer, compare it to the background module, and find the error or the part of the code where the answers differ. The tracer is then the meta-interpreter, an arbitrarily complex program that evaluates another program, as in the original AD by Murray (1987). The predicate returned as the solution would be the misconception in the student's understanding. However, this idea did not work either as the tracer needs to know what kind of input to expect. As the next list element of an equation list always differs (it could be an atom, a molecule, an electron, or a sign), such an approach is hardly feasible as many different predicates would be needed which contradicts the idea of a tracer. Recall that a tracer aims to unify a variable with some part of the backend, so the input (the formatting of that variable) is crucial. Let me explain this in more detail using an example: Assigning oxidation numbers may be considered the most difficult step as it comprises the most rules and many students struggle with it. Furthermore, it is likely to be the most promising in testing a tracing approach as it is more rule-based than the splitting of molecules where several exceptions occur. Hence, I intended to implement a more fine-grained error diagnosis system for oxidation numbers using a tracer. The great problem was, however, that a "proper" tracing as in classical AD did not work for my use case because of the brackets around molecules. As tracing needs matching or unification, it does not work properly with molecules as they are lists. Hence, several predicates would be needed to make Prolog expect lists as input. As molecules can occur anywhere within a formula, such a predicate would be very extensive and slightly hardcoded: Every possible equation must be considered as input. This is not only computationally expensive but contradicts the idea of AD in that the tracer should iterate over any type of input instead of providing every possible input, so there is hardly any tracing left to do. At this point, just as with the analogies, tracing seems superfluous: Any sensible tracer would have to be so wide and cover so many input cases that it is not really a tracer anymore but simply a predicate looking for differences in student and expert solution—and this is exactly what I implemented.

A second reason why I decided against Zinn's (2014) original implementation of AD in an ITS is that his approach is not step-based. A simple tracer may be less computationally demanding than a step-based approach, but, again, such a simple tracer is not realizable with the different input types anyway. A very complex tracer is also computationally demanding but not step-based. As research suggests that step-based approaches are superior to other approaches (VanLehn, 2011), I decided to follow the suggested guideline of implementing a feedback system with a step-based granularity. Hence, as with the analogies, I discarded my original idea of strictly following AD. Instead, the whole system is separated into steps. After each step, the solutions are compared using the oracle and the `==/2`-predicate and, if erroneous, an analogous solution to that sub-step is presented. This arguably still follows the idea behind AD, which is to compare the student's answer with the answer of the expert module and, if erroneous, track that error down (in my case, locate it in a sub-step). However, it no longer involves a tracer. This seems like a great shortcoming as it seemingly does not allow for precise localization of the error within a specific predicate. Some more fine-grained error detection is undoubtedly useful. Again, however, such a localization is also not possible with a tracer because of the formatting. Furthermore, some thoughts have been put into possible error detection without a tracer by adding the `same_list/2`-predicate. It allows one to localize exactly where the (first) error occurred. For example, by applying `same_list/2` we know which element has been assigned an incorrect oxidation number. Now a tracer could be used to check which rule was responsible for assigning its correct oxidation number. But my approach here is shifting the precise error location away from the backend module. The two outputs, one from the user, and the other from the backend, could be compared in the Python module to look for their disagreement. This error could then be highlighted or additional information on that error could be provided. Note that this gives precise error feedback but does not need a tracer: Simply highlighting the incorrect part of the answer is perfectly possible using the `same_list/2`-predicate and some extension in the frontend (see below). Providing general information on the sub-step and its typical errors also works without knowing exactly which rule failed,[2] so a tracer is not necessary to provide detailed feedback.

Instead, additional test cases are needed to know if this error feedback works. For example, when setting up the partial equations, you could check whether molecules have been split correctly, whether the oxidation numbers have been taken as the charge, whether the electrons have been written on the correct side of the equation, and so on. I already added such typical error cases per sub-step in the Test Suite. These can be used to test the individualized error feedback discussed before.

[2] Again, it is questionable if one precise rule could even be found given the list constraints and the fact that it is often not possible to locate a chemical misconception in one rule precisely.

I will now discuss the findings of the interviews, especially on the usability of the frontend. First, conducting both interviews went well, although the timing was slightly off as it took both participants longer than expected to answer an equation in the frontend. This is partly due to the current input format, but also due to them trying out many different errors or inputs. Hence, only two reactions each could be solved, but both stated that answering only those two was sufficient to try out the system. The open questions and the rest of the interview took roughly as much time as anticipated. Many interesting ideas were voiced during these interviews that align with my thoughts on advantages, current shortcomings, or future enhancements of the system.

For example, both participants rated the division into easy and difficult formulas as sensible, while at the same time neither mentioned any (significant) differences between solving them. This speaks in favor of the system, as the handling appears equally intuitive in both cases. In general, both praised the design and layout of the system. The step-by-step procedure and the immediate feedback were particularly helpful. This confirms research findings on tutoring systems, highlighting immediate feedback as helpful and emphasizing the superiority of step-based ITS.

The analogies as a type of feedback are also confirmed here, they were helpful and understandable. Both mentioned the similar structure between the analogy and the formula, so the selection of analogies seems sensible, as it was precisely the similarity of the structure that should be recognized and used. Both compared the analogy with the formula while thinking aloud: $P1$ realized that the hydrogen is bound to the electron in the analogy, which probably was her mistake. $P2$ recognized the correct representation of the charges within a molecule using the analogous example.

At the same time, both criticized the lack of a direct comparison between the analogy and their solution. This should be added (see below), as the very idea of analogies is for students to directly compare their incorrect solution and the analogy to deduce the correct solution. Furthermore, the processing time should be shortened. None of the interviewees mentioned this explicitly, but working on a formula took about half an hour. This is too long even if you consider the time it takes to get used to the system. It is doubtful that students would use such a lengthy tool.

Interestingly, there was disagreement about providing extra help along with the analogies. While $P1$ would have liked to see more individual error feedback immediately, $P2$ recommends only displaying analogies first. Students should first observe and find an idea themselves, and only if they answered incorrectly twice should further help be displayed. In my observation, this fits their level of prior knowledge. $P2$ was always able to immediately read the correct feedback from the analogies. For example, she immediately recognized that the molecular charge in the analog example was different and that this was probably her mistake. $P1$, with less prior knowledge of chemistry, was often unable to make such transfer steps. From this, one could cautiously conclude that the less prior knowledge a person has, the more accurate the error feedback should be. However, further and larger studies should be carried out to fully decide whether and how the analogies should be extended as feedback (see below).

Overall, there are some important shortcomings in the system, both in front- and backend. Nevertheless, the system has been rated positively in the interviews, both regarding its usability and the realization of the use case. Most importantly, it has been stated that, with a few alterations, the system can be used in schools to help students overcome their problems and misconceptions when solving chemical equations. Considering the shortcomings, this implementation thereby fulfilled its intention of providing a prototypical, working tutoring system in the much-needed domain of chemical equations that is straightforward for students to understand and use.

6.4 Future Work

Drawing on the previously discussed findings, several enhancements to the system are suggested here: Firstly, more testing and a quantitative evaluation such as empirical studies are needed before the ITS can be used. Regarding the impact of the created analogies, this could be done by comparing the analogies provided by the system to analogies suggested by a human tutor. For example, Myneni et al. (2013) conducted experiments including a usability questionnaire and the data from the procedure: In a pre-test, students were asked to answer 18 questions (multiple-choice) related to their systems (so the system could measure their knowledge and checked if they showed any of the misconceptions). Then they solved the problems before answering the same 18 questions in a different order in a post-test. A similar procedure could be useful for my tutoring system as well. Another aspect could be to explicitly ask for positive or negative aspects of the tutoring system in a greater interview study. This may again give hints on its usability. Yet another useful evaluation aspect is to check for comprehensibility. However, this is quite complex, as comprehensibility is difficult to measure. It would be necessary to operationalize what is meant by comprehensibility and then carry out (experimental) studies to check if one type of instruction is easier to grasp than another. Nevertheless, this may be fruitful as the feedback of the conducted qualitative analysis suggests that the instructions are not always clear.

As noted before, balancing atoms is recommended as a test step to check that the number of each atom in the educt equals the number of the same atom in the product. However, it is not implemented in the expert module. If such a step becomes necessary, it should be added. Note that as this is intended as a test step only, it would not balance out any atoms (this is part of the previous steps), but simply throw an error if any atoms are not yet balanced. Using `findall/3`, all atoms that are members of the equations are found and put into a list. Next, by iterating over this list, it is checked that the number of each element in the educt is the same as the difference in the product. This is reported as an error message, if it fails on one predicate. Similarly, reducing could be realized by iterating over all coefficients using `findall/3` and building the greatest common denominator. If each coefficient is then divided by this greatest common denominator, the equations should be reduced. Alternatively, it may be fruitful to implement such a check in the frontend. For example, before moving on to the last step, a table listing all elements and their numbers within the formula could be presented so the student can check if they are indeed balanced.

The input formula of the frontend has to be adapted. In the case of an equation, the usual chemical notion, cf. of

$$H_2 + O \rightarrow H_2O \tag{6.3}$$

should be typed in. The frontend could provide input masks to ease the handling of the input further. As an example, the results of the oxidation numbers could simply be typed into the provided boxes. The input of these boxes is then parsed to be readable for Prolog and compared against the result of the expert module. Also, it could be ensured that students can easily type in numbers or $+/-$ symbols as sub- or superscripts to allow for indices and charges. Lastly, the input misses marking the passive elements in an equation in grey. Recall that sometimes (for example if metal oxides and metals react) elements are part of a reaction although they are neither part of the oxidation nor the reduction. These elements are simply listed in the half-reactions but need not be balanced. To detect these elements better, they should be marked in grey in the input formula.

Based on a better input format, a more thorough checking of formal errors could be added to control for formal mistakes such as mistyping or incorrect commas. Such an error check should ensure that an answer is not considered false just because the ordering does not match the expected format, for example, if electrons are not placed at the end of an equation. This way, only content errors are identified in the backend and only in those cases, an analogy is given.

Moreover, analogy and the incorrect solution should be presented next to each other to decrease the mental load of correctly remembering one's previous solution. This will help students to focus on the comparison even more, so the analogies will be even more useful.

To shorten the execution time of solving redox reactions, it is fruitful to accept the solution of the previous sub-step by clicking on it if the solution is also correct for the current sub-step. Some sub-steps may also be left out for the easy equations.

The information page and the tutorial should be added to the front and the instructions in the *next_step.html*-template should be more precise. Additionally, information on the sub-steps could be available in the frontend while solving the equation. For example, a button providing basic information on the current step could be pressed if a student is unsure what to do.

But how to present such additional feedback? As a first idea, the granularity of the feedback could be linked to the mode: If you are using the tool for the first time, additional help, such as highlighting, could be displayed alongside the "pure" analogy. In expert mode, however, only the analogy is displayed. Alternatively, only the analogy could be displayed for the first incorrect solution. If this is a sufficient explanation, the person will solve the step correctly. For a second incorrect solution, the error in the analogy could be highlighted, for a third incorrect solution, additional information could be displayed, etc. The granularity here would not be based on the mode, but rather on the number of incorrect answers.

Some chemical extensions are also useful. Firstly, the scope of redox reactions could be broadened to include more equations. But redox reactions are typically formulated as text tasks in schools, such as "Formulate the reaction equation for the decomposition of silver oxide and name the particle that is oxidized and the one that is reduced" (Mihlan and Töhl-Borsdorf, 2009, p. 104; translated from German) or "Formulate the reaction equation for the extraction of aluminum from a melt of alumina by electrolysis. Identify oxidation and reduction" (Mihlan and Töhl-Borsdorf, 2009, p. 108; translated from German). Hence, it may be even more useful to expand the system to solve such text tasks as well, especially given the abundance of text exercises in chemistry textbooks.

Other chemical reactions following the same sub-steps, including acid-base reactions with protons. These are not redox reactions since protons, not electrons, are transferred. However, they are very similar in their structuring and the underlying chemical principles. A different element is transferred, but the grid remains the same. Whole molecules are exchanged in other reactions following the same structuring as redox reactions (mail with ChEx, 18.03.24). One could therefore try to extend the existing system to include such chemical reactions.

Lastly, the teaching perspective could be broadened to include other aspects or levels of chemistry, especially for grasping the trichotomy underlying redox reactions and, thereby, an even greater level of abstraction (Taskin and Bernholt, 2012). This could be done by explaining the underlying submicroscopic chemical level at least to the information page or as part of the information pop-ups.

Conclusion 7

Drawing on work by Zinn (2014) and Zeller (2015) this master thesis aimed to answer if the built tutoring system is easy to understand and use and if the system can help students solve chemical equations, mainly redox reactions, in the future. This question arose out of the need for chemistry-related tutoring systems. Although tutoring systems are common in other fields such as medicine or mathematics and are very useful, especially in STEM-related fields, there are hardly any chemical ITS in practice and none on chemical reactions. But as many students struggle with redox reactions in particular and fail to grasp the underlying logic and rules inherent to solving these reactions (Taskin and Bernholt, 2012), it suggested itself as a field of application. To do so, I implemented the backend expert module in Prolog which correctly solves a redox reaction it receives as input. This expert answer, roughly following the idea of AD, is called when checking if the student's answer is correct. The system is step-based, so the procedure of solving redox reactions is split into sensible sub-steps. The provided view presents the current step to users and receives their answers. The controller then redirects these answers and calls the student module in the backend to check them. If the expert and student answer match, the controller ensures the next step is presented in the frontend. Otherwise, the feedback module chooses a structurally analogous example, which is presented to users and hopefully helps them detect their errors and solve the step correctly.

Upon evaluating the implemented system in an interview analysis, it reportedly works well and reliably while being easy to understand, so it can be used in classroom settings in the future. It is fruitful to refine some instructions, add individualized error feedback, and some more information on the submicroscopic chemical level in the future to fully grasp the potential of both analogies and this tutoring system.

M. Stümke, *Identifying and Overcoming Student's Misconceptions in Chemistry*, BestMasters, https://doi.org/10.1007/978-3-658-51032-9_7

Bibliography

Abe, Y. (2019). Use of analogy in the development of intercultural competence. *Philosophies, 4*(2), 25. https://doi.org/10.3390/philosophies4020025

Almasri, A., Ahmed, A., Almasri, N., Abu Sultan, Y. S., Mahmoud, A. Y., Zaqout, I. S., Akkila, A. N., & Abu-Naser, S. S. (2019). Intelligent tutoring systems survey for the period 2000–2018.

Anderson, J. R., Boyle, C. F., & Reiser, B. J. (1985). Intelligent tutoring systems. *Science, 228*(4698), 456–462. https://doi.org/10.1126/science.228.4698.456

Bandura, A. (1977). Self-efficacy: Toward a unifying theory of behavioral change. *Psychological Review, 84*(2), 191–215. https://doi.org/10.1037/0033-295x.84.2.191

Bernholt, S. (2021). Symbol—Modell—Kontext: Anforderungen und Schwierigkeiten beim Erlernen der chemischen Formelsprache. *Naturwissenschaften im Unterricht- Chemie, 32*(181), 2–9.

Bounhas, M., & Prade, H. (2023). Analogy-based classifiers: An improved algorithmexploiting competent data pairs. *International Journal of Approximate Reasoning, 158*, 108923. https://doi.org/10.1016/j.ijar.2023.108923

Brown, J., & Burton, R. (1978). Diagnostic models for procedural bugs in basic mathematical skills. *Cognitive Science, 2*(2), 155–192. https://doi.org/10.1016/s0364-0213(78)80004-4

Brown, S., & Salter, S. (2010). Analogies in science and science teaching. *Advances in Physiology Education, 34*(4), 167–169. https://doi.org/10.1152/advan.00022.2010

Caballero, R., Riesco, A., & Silva, J. (2017). A survey of algorithmic debugging. *ACM Computing Surveys, 50*(4), 1–35. https://doi.org/10.1145/3106740

Cabubas, D. N., & Mustacisa-Lacaba, M. (2023). Utilization of pictorial analogy and colored sticks in teaching chemical equation. *Jurnal Penelitian dan Pembelajaran IPA, 9*(1), 15. https://doi.org/10.30870/jppi.v9i1.18873

Coll, R. K. (2004). The role of models, mental models and analogies in chemistry teaching. In *Science & technology education library* (pp. 65–77). Springer-Verlag. https://doi.org/10.1007/1-4020-3830-5_6

Corbett, A. T., Koedinger, K. R., & Anderson, J. R. (1997). Intelligent tutoring systems. In *Handbook of human-computer interaction* (pp. 849–874). Elsevier. https://doi.org/10.1016/b978-044481862-1.50103-5

M. Stümke, *Identifying and Overcoming Student's Misconceptions in Chemistry*, BestMasters, https://doi.org/10.1007/978-3-658-51032-9

Crippen, K. J., & Earl, B. L. (2007). The impact of web-based worked examples and self-explanation on performance, problem solving, and self-efficacy. *Computers & Education, 49*(3), 809–821. https://doi.org/10.1016/j.compedu.2005.11.018

Flemming, T. M., Thompson, R. K. R., & Fagot, J. (2013). Baboons, like humans, solve analogy by categorical abstraction of relations. *Animal Cognition, 16*(3), 519–524. https://doi.org/10.1007/s10071-013-0596-0

Forbus, K. D. G., Everett, D., & J. O. Wu, M. Towards a computational model of evaluating and using analogical inferences. In: In *Proc. of the 19th annual conference of the cognitive science society*. 1997, 229–234.

Freeman, E., Sierra, K., & Bates, B. (2008). Head first design patterns. O'Reilly.

Gentner, D. (1983). Structure-mapping: A theoretical framework for analogy. *Cognitive Science, 7*(2), 155–170. https://doi.org/10.1016/s0364-0213(83)80009-3

Gentner, D. (2003). Why we're so smart. In S. Gentner D.; Goldin-Meadow (Ed.). Cambridge, MA: MIT Press.

Gentner, D., Loewenstein, J., & Thompson, L. (2003). Learning and transfer: A general role for analogical encoding. *Journal of Educational Psychology, 95*(2), 393–408. https://doi.org/10.1037/0022-0663.95.2.393

Gentner, D., & Smith, L. A. (2013). *Analogical learning and reasoning*. Oxford University Press. https://doi.org/10.1093/oxfordhb/9780195376746.013.0042

Gerjets, P., Scheiter, K., & Catrambone, R. (2006). Can learning from molar and modular worked examples be enhanced by providing instructional explanations and prompting self-explanations? *Learning and Instruction, 16*(2), 104–121. https://doi.org/10.1016/j.learninstruc.2006.02.007

Gick, M. L., & Holyoak, K. J. (1980). Analogical problem solving. *Cognitive Psychology, 12*(3), 306–355. https://doi.org/10.1016/0010-0285(80)90013-4

Goes, L. F., Nogueira, K. S. C., & Fernandez, C. (2020). Limitations of teaching and learning redox: A systematic review. *Problems of Education in the 21st Century, 78*(5), 698–718. https://doi.org/10.33225/pec/20.78.698

Hansen, J. (2022). *Trainingsmöglichkeiten von Formelsprache im Anfangsunterricht* [Master's thesis, Christian-Albrechts-Universität zu Kiel].

Harsley, R., Green, N., Alizadeh, M., Acharya, S., Fossati, D., Di Eugenio, B., & Al-Zoubi, O. (2016). Incorporating analogies and worked out examples as pedagogical strategies in a computer science tutoring system. *Proceedings of the 47th ACM Technical Symposium on Computing Science Education*. https://doi.org/10.1145/2839509.2844637

Heuer, S. (2009). *Kommunikation im Chemieunterricht: Die chemische Formelsprache als Element von Kompetenzentwicklung* [Master's thesis, Universität Oldenburg].

Insa, D., & Silva, J. (2018). Algorithmic debugging generalized. *Journal of Logical and Algebraic Methods in Programming, 97*, 85–104. https://doi.org/10.1016/j.jlamp.2018.02.003

Koch, A., & Westhoff, K. (2012). *Task-Analysis-Tools (TATоо)—Schritt für Schritt Unterstützung zur erfolgreichen Anforderungsanalyse*. Pabst Science Publishers.

Kozma, R., & Russell, J. (2005). Multimedia learning of chemistry. *The Cambridge Handbook of Multimedia Learning*, 409–428.

Leifi-Chemie. (2024). *Oxidationszahlen*. Joachim Herz Stiftung. https://www.leifichemie.de/redoxreaktionen-und-elektrochemie/redoxreaktionen-als-elektronenuebertragungen/grundwissen/oxidationszahlen

Leifi-Physik. (2023). *Einführung Redoxreaktionen*. Joachim Herz Stiftung. https://www.leifichemie.de/allgemeine-chemie/einfuehrung-redoxreaktionen

Lola, A., & Tzetzis, G. (2020). Analogy versus explicit and implicit learning of a volleyball skill for novices: The effect on motor performance and self-efficacy. *Journal of Physical Education and Sport, 2020*(05). https://doi.org/10.7752/jpes.2020.05339

Lulis, E., Evens, M., & Michael, J. (2004). Implementing analogies in an electronic tutoring system. In *Lecture notes in computer science* (pp. 751–761). Springer Berlin Heidelberg. https://doi.org/10.1007/978-3-540-30139-4_71

Mihlan, M., & Töhl-Borsdorf, J. (2009). *Elemente chemie 9|10* (B. Sommer & A. Tompert, Eds.). Ernst Klett Verlag.

Mousavinasab, E., Zarifsanaiey, N., R. Niakan Kalhori, S., Rakhshan, M., Keikha, L., & Ghazi Saeedi, M. (2018). Intelligent tutoring systems: A systematic review of characteristics, applications, and evaluation methods. *Interactive Learning Environments, 29*(1), 142–163. https://doi.org/10.1080/10494820.2018.1558257

Murray, W. R. (1987). Automatic program debugging for intelligent tutoring systems. *Computational Intelligence, 3*(1), 1–16. https://doi.org/10.1111/j.1467-8640.1987.tb00169.x

Myneni, L. S., Narayanan, N. H., Rebello, S., Rouinfar, A., & Pumtambekar, S. (2013). An interactive and intelligent learning system for physics education. *IEEE Transactions on Learning Technologies, 6*(3), 228–239. https://doi.org/10.1109/tlt.2013.26

Novick, L. R., & Holyoak, K. J. (1991). Mathematical problem solving by analogy. *Journal of Experimental Psychology: Learning, Memory, and Cognition, 17*(3), 398–415. https://doi.org/10.1037/0278-7393.17.3.398

Orgill, M., & Bodner, G. (2004). What research tells us about using analogies to teach chemistry. *Chem. Educ. Res. Pract., 5*(1), 15–32. https://doi.org/10.1039/b3rp90028b

Paun, N. (2015). Chemlogic: A logic programming computer chemistry system.

Pritchard, T. (2018). Analogical cognition: An insight into word meaning. *Review of Philosophy and Psychology, 10*(3), 587–607. https://doi.org/10.1007/s13164-018-0419-y

Rau, M. A., Michaelis, J. E., & Fay, N. (2015). Connection making between multiple graphical representations: A multi-methods approach for domain-specific grounding of an intelligent tutoring system for chemistry. *Computers amp; Education, 82*, 460–485. https://doi.org/10.1016/j.compedu.2014.12.009

Rowling, J. K. (1997). *Harry potter and the philosopher's stone*. Bloomsbury.

Sarantopoulos, P., & Tsaparlis, G. (2004). Analogies in chemistry teaching as a means of attainment of cognitive and affective objectives: A longitudinal study in a naturalistic setting, using analogies with a strong social content. *Chem. Educ. Res. Pract., 5*(1), 33–50. https://doi.org/10.1039/b3rp90029k

Shapiro, E. (1982). *Algorithmic program debugging* [Doctoral dissertation, Yale University].

Sterling, L., & Shapiro, E. (1994). *The art of prolog: Advanced programming techniques*. MIT Press.

Stieff, M. (2005). Connected chemistry—a novel modeling environment for the chemistry classroom. *Journal of Chemical Education, 82*(3), 489. https://doi.org/10.1021/ed082p489

Taskin, V., & Bernholt, S. (2012). Students' understanding of chemical formulae: A review of empirical research. *International Journal of Science Education, 36*(1), 157–185. https://doi.org/10.1080/09500693.2012.744492

Thielsch, M. T., Lenzner, T., & Melles, T. (2012). Wie gestalte ich gute Items und Interviewfragen? In M. T. Thielsch & T. Brandenburg (Eds.), *Praxis der Wirtschaftspsychologie. Themen und Fallbeispiele für Studium und Anwendung* (pp. 221–240).

VanLehn, K. (2011). The relative effectiveness of human tutoring, intelligent tutoring systems, and other tutoring systems. *Educational Psychologist, 46*(4), 197–221. https://doi.org/10.1080/00461520.2011.611369

Warren, D. S. (2023). Introduction to prolog. *In Lecture notes in computer science* (pp. 3–19). Springer. https://doi.org/10.1007/978-3-031-35254-6_1

Wiese, E., Konderding, U., & Schmid, U. (2008). Mapping and inference in analogical problem solving—as much as needed or as much as possible? *Proceedings of the Annual Meeting of the Cognitive Science Society, 30.*

Wu, H.-K., Krajcik, J. S., & Soloway, E. (2001). Promoting understanding of chemical representations: Students' use of a visualization tool in the classroom. *Journal of Research in Science Teaching, 38*(7), 821–842. https://doi.org/10.1002/tea.1033

Zeller, C., & Schmid, U. (2017). Automatic generation of analogous problems to help resolving misconceptions in an intelligent tutor system for written subtraction [https://fis.uni-bamberg.de/handle/uniba/41882], 10. Vortrag, gehalten auf: 24th International Conference on Case Based Reasoning, Atlanta, GA, 31th October-2nd November 2016.

Zeller, C. (2015). *Automatische Erzeugung analoger Beispiele aus Debugging-Traces: Konstruktion von Rückmeldungen für ein Intelligentes Tutor-System zur schriftlichen Subtraktion* [Master's thesis, Otto-Friedrich-Universität Bamberg].

Zinn, C. (2011). Algorithmic debugging to support cognitive diagnosis in tutoring systems. In *Lecture notes in computer science* (pp. 357–368). Springer Berlin Heidelberg. https://doi.org/10.1007/978-3-642-24455-1_35

Zinn, C. (2014). Algorithmic debugging and literate programming to generate feedback in intelligent tutoring systems. In *Ki 2014: Advances in artificial intelligence* (pp. 37–48). Springer International Publishing. https://doi.org/10.1007/978-3-319-11206-0_4

The manufacturer's authorised representative in the EU is Springer Nature Customer Service Centre GmbH, Europaplatz 3, 69115 Heidelberg, Germany. If you have any concerns regarding our products, please contact ProductSafety@springernature.com

Printed and bound by CPI Group (UK) Ltd, Croydon, CR0 4YY

05/10/2026

02228034-0002